LIVEBEARING
FISHES

LIVEBEARING FISHES

A GUIDE TO THEIR AQUARIUM CARE, —— BIOLOGY AND CLASSIFICATION ——

JOHN DAWES

BLANDFORD

Paperback first published in the UK in 1995 by
Cassell plc, Wellington House, 125 Strand,
London WC2R 0BB
Hardback published 1991

Distributed in the USA by Sterling Publishing Co. Inc.,
387 Park Avenue South, New York, NY 10016-8810

Distributed in Australia by Capricorn Link (Australia) Pty. Ltd.,
2/13 Carrington Road, Castle Hill, NSW 2154

British Library Cataloguing-in-Publication Data
Dawes, John A. (John Albert) 1945-
  Livebearing fishes: a guide to their aquarium care,
  biology and classification.
  1. Aquariums. Livebearing fish. Breeding. Breeding & care.
  I. Title
639.3753

ISBN 0-7137-2592-3

Typeset by Fakenham Photosetting Ltd, Fakenham, Norfolk

Printed and bound in Spain by Graficromo

Contents

Part II Aquarium Care

Part III   Selected Species and Varieties

Catalogue of Selected Species and Varieties   **128**

ACKNOWLEDGEMENTS

I would like to extend sincere thanks to the following individuals and institutions without whose generous assistance this book would not have been possible.

Among leading UK aquarists, Dennis Barrett, Derek and Pat Lambert, Colin Vernon and Wilf Blundell provided me with illustrations, data, advice, literature and unselfish, unlimited helpings of their invaluable time. On the US front, Ross Socolof was equally helpful.

Jim Chambers of the Fish Section at the British Museum (Natural History) and Dr. Peter Miller of the Zoology Department at Bristol University helped me trace vital research papers and pointed me in numerous fruitful directions.

Jaap Jan de Greef supplied me with valuable first-hand information regarding *Xenodexia*, while Mary Bailey translated essential German data expertly and in double-quick time.

Aquarian Fish Foods, Interpet Ltd., the Centre for Electron Optics at Bath University, Dr. Harry Grier and the Florida Tropical Fish Farms Association granted me permission to use some of their excellent illustrations.

I am indebted to them all.

John Dawes

Preface

Every aquarist is familiar with the term 'livebearer'. Some of the most popular and best-loved aquarium fish, like Guppies, Swordtails, Platies and Mollies, belong to this group. These fish are so widely available and well known that no 'first' aquarium can really be considered complete without at least a few livebearing representatives.

Yet, these popular fish represent only a small part of a very substantial group, consisting of nearly 950 species of fish that could be regarded as livebearers.

Livebearing is scattered among some 40 families (and approximately 420 species) of so-called cartilaginous fish, with representatives among the sharks, rays and dogfish. Bony fish have livebearing representatives in 14 families (and around 510 species) – ranging from the universally-known examples cited above to some less-known (and more surprising) ones. These include the Eelpouts, Baikal Oilfish, some species of Scorpionfish, the Surfperches, the Kelpfish and several others. The most interesting is, almost certainly, that world-famous living fossil, the Coelacanth.

I have concentrated in this book largely on those species that could be termed loosely 'aquarium livebearers'. This categorisation raises problems, of course. What, for instance, is an 'aquarium' species? And, even if we could define the term, what *is* a livebearer?

There are no clear-cut answers to these deceptively simple questions. I have, nevertheless, attempted to formulate a working definition of what constitutes an aquarium livebearer. This is an imperfect definition that is bound to change with time as more and more hitherto 'non-aquarium' species are kept and bred by aquarists.

I have tried also to present a comprehensive review of the concept of livebearing. This is aimed primarily at all those who are interested in finding out more about those fish we call livebearers than is possible within the confines of a more general aquarium book.

In addition, I have taken a look at the biology of livebearers, their distribution and aquarium care, providing in the process some guidelines on aquarium layouts, provision of adequate breeding facilities, nutrition, health and other aspects of captive maintenance.

In the final section, I have compiled an extensive and up-to-date selection of all the well-known species and varieties, plus many of the lesser-known ones, including a few examples that some readers may find surprising.

There are no fixed boundaries to the aquarist's hobby and anyone who likes a challenge, has an open mind and hungers for knowledge will find these fuzzy boundaries uniquely attractive and exciting.

I have enjoyed myself enormously during the preparation of this book and have learned a great deal in the process. It is my sincere hope that some of this enjoyment will come through in the pages that follow and that readers will end up as I have done, with more questions than I had when I started!

John Dawes

Part I
Biology

What is a Livebearer?

A Guppy is a livebearer; a Goldfish is an egglayer. Few people would dispute this. After all, Guppies don't lay eggs; they give birth to fully formed fry. On the other hand, Goldfish, along with a whole host of other species, lay eggs that are fertilised as soon as they leave the female's body, subsequently undergoing development and hatching in the water itself.

It seems easy enough: livebearers produce live young, while egglayers, true to their name, lay eggs. But in fact things are not quite as simple as this apparently foolproof two-way division might suggest – the deceptively uncomplicated question that heads this section has a rather convoluted and complex answer. We may even have to reach the conclusion that it is impossible to define exactly, concisely and without qualification just what a livebearer is!

Live Young versus Egg Production

Take the term 'livebearer' itself. What does it really mean? If we say that it means the production of live young, then we are implying that egglayers do *not* produce live young. But the existence of billions of living egglayer fry shows that this is untrue, so a definition of livebearing based on this criterion alone cannot be valid.

If we turn to other aspects of a potential definition, new flaws appear. For instance, if we choose egg production as the chief criterion, the argument falls flat almost before it gets going. *All* fish, be they livebearers or egglayers, produce eggs. The difference between them lies only in what happens to the eggs once they have been produced.

Internal versus External Fertilisation

Because fish live in an aqueous medium, they, unlike land-based organisms, can opt either for internal or for external fertilisation of their eggs. All livebearers employ internal fertilisation. It would therefore seem logical to use this criterion as the deciding factor in separating livebearers from egglayers. However, while it may be true that all livebearers use *internal* fertilisation, it is equally true to say that not all egglayers use *external* fertilisation.

Plate 1 A Guppy giving birth. (*Harry Grier*)

Plate 2 BELOW LEFT This is a livebearer egg. Females of this particular species (*Ameca splendens*) not only ovulate, as do mammals, but also nourish their embryos during development. (*Michael King*)

Plate 3 BELOW RIGHT A pair of Amarillos (*Girardinichthys viviparus*) a split second before mating. All livebearers need to make contact for successful insemination to occur. (*Dennis Barrett*)

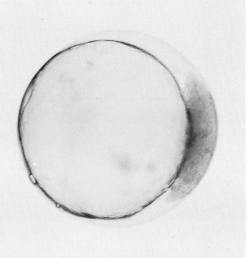

There are numerous examples of this, mainly among the cartilaginous fishes (Chondrichthyes, the sharks and rays) but also, to a much lesser extent, among the bony fishes (Osteichthyes).

Among the Characidae (the family that includes those delightful and highly popular fish, the Neon and Cardinal Tetras, as well as the Piranha), there is one species, *Corynopoma riisei* (the Swordtail Characin), which employs internal fertilisation, followed by egg deposition.

Among the Oryziatidae, *Oryzias latipes* (the Medaka, Rice Fish or Geisha Girl) usually employs external fertilisation but can, on occasion, fertilise its eggs internally.

Then there are at least two Killifish – *Cynolebias (Cynopoecilus) melano-taenia* and *Cynolebias (Campellolebias) brucei* – in which the males possess anal fin modifications believed to be for internal fertilisation purposes. Female *Cynolebias brucei* kept on their own after sharing a tank with males, have been known to lay fertilised eggs.

Another Killifish, *Rivulus marmoratus*, has been found to exhibit self-fertilising, synchronous/functional hermaphroditism (both sexes present in the same individual at the same time). In this case, a single individual is capable of fertilising its own eggs internally, later depositing them to complete their development externally.

Two further examples deserve mention. *Horaichthys setnai*, the sole representative of its genus and family (Horaichthyidae), is a highly special-ised fresh- and brackish-water fish found in coastal western India. Males have an elaborately developed split anal fin which looks superficially like the gonopodium of the best-known livebearers and which, as in these fish, is used as a copulatory organ. The internally fertilised eggs are then laid by the females and complete their development in typical egglayer fashion. In the southeast Asian fresh- and brackish-water Phallostetheids, the copula-tory organ (priapium) is even more elaborate than that of *Horaichthys*. So, clearly, internal fertilisation of eggs cannot be used as a fool proof diagnos-tic feature separating livebearers from egglayers.

The argument is given a further, interesting twist by *Tomeurus gracilis*, a member of the same family as the Guppies, Mollies, Swordtails and Pla-ties, all of which have always been regarded as classic livebearers. *Tomeur-us*, too, employs internal fertilisation with subsequent laying of the eggs

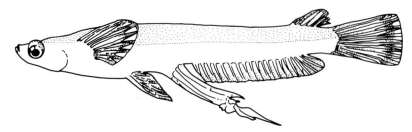

Fig. 1 In *Horaichthys setnai*, the eggs are fertilised internally but are then released by the female. (Based on A. Wheeler, *World Encyclopaedia of Fishes*, Macdonald 1985)

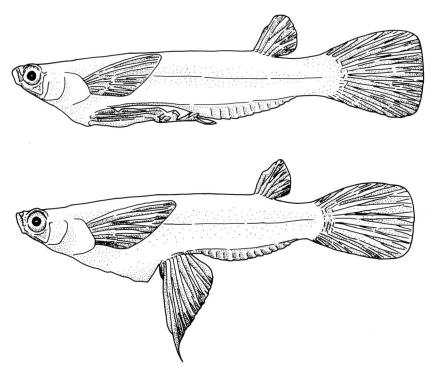

Fig. 2 *Tomeurus gracilis* is a unique Poeciliid in that it employs internal fertilisation with subsequent deposition of eggs. (Based on D. E. Rosen and K. D. Kallman, *Quart. Journ. Fla. Acad. Sci.*, **22** (3), 1959)

and, at first glance, appears to bear a remarkable resemblance to *Horaichthys*. However, its copulatory organ is a true gonopodium which, as in its close relatives, is a highly modified anal fin, but it is not 'split' as in *Horaichthys*.

Internal Fertilisation and Egg Retention

If a single criterion cannot be used to formulate a definition of a livebearer, might a combination of criteria yield better results?

Let us look at some examples of fish that combine internal fertilisation with egg retention up to the moment of birth.

That most famous of all living fossils, the Coelacanth (*Latimeria chalumnae*), may not be quite as primitive (at least, in some respects) as many people believe it to be, for it shares a rather advanced characteristic with many of our best-known livebearers – internal fertilisation *and* retention of eggs up to the moment of birth.

Latimeria produces eggs the size of a large orange which are fertilised internally, once ovulation has taken place. This is followed by hatching inside the female and subsequent development of the embryos within the oviduct. During gestation – the period between fertilisation and parturition – it is estimated that the weight of the developing embryos can change anywhere from −2% (slight loss of weight) to as much as +43% (considerable increase in weight). Lack of adequate research material – after all, not that many Coelacanths have been caught and studied and, of these, even fewer have been pregnant females – is the main reason for this apparently wide range of estimates. Even so, we have enough data to conclude that developing embryos receive some nourishment during their 10–13 months' gestation, in addition to that supplied by their yolk sacs. Whether that extra nourishment comes from secretions produced by the female, or through the consumption of histotrophe (cellular debris, or breakdown products from dying eggs and embryos), or even through the consumption of whole eggs by the growing foetuses, is still under debate.

Many aquarists, particularly marine hobbyists, are familiar with *Zoarces viviparus* – the fish commonly known as the Livebearing Blenny, although in fact it is not a Blenny at all but an Eelpout belonging to the family

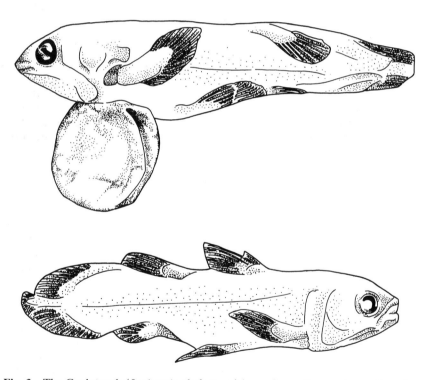

Fig. 3 The Coelacanth (*Latimeria chalumnae*) is, perhaps, the most arresting of all the livebearers. (Based on E. K. Balon, in G. W. Potts and R. J. Wooton (eds), *Fish Reproduction: Strategies and Tactics*, Academic Press 1984)

Zoarcidae. Like the Coelacanth, the Livebearing Blenny exhibits both internal fertilisation and internal retention of the eggs up to the moment of birth. What is more, the embryos can increase by more than 1000% in weight during their 3½- to 4-month gestation!

I have given only two examples of fish that use internal fertilisation and egg retention and don't quite fit our traditional image of a livebearer, but numerous similar examples could be cited, ranging from Blue Sharks and many Scorpionfish (Scorpoaenidae) to Halfbeaks.

Degrees of Livebearing

All biological categorisations and classifications are artificial constructs imposed by us on a naturally existing variable spectrum of possibilities. For their part, the fish do what comes naturally – they merely provide their own biological solutions to the matter of species survival, each modified and finely tuned by evolutionary pressures through countless generations. As a result, there are often no sharply defined limits between one reproductive strategy and another.

Even within livebearing itself, intergradations occur. At one extreme, there is the form of livebearing exhibited by Guppies, Mollies and their close relatives, while, at the other, there is the 'truer' livebearing strategy of Goodeids and others.

In the former category, the anal fin of the males is highly modified into a gonopodium, widely regarded as *the* main external characteristic separating livebearers from all other types of fish. Further, the precise arrangement of the spines, hooks, claws and blades that are present around the extremity of the gonopodium is unique for each species. The fin rays most affected by these elaborations are the third, fourth and fifth. The other rays, while appearing insignificant, are nevertheless absolutely essential in helping to provide the grooved channel through which the spermatozeugmata, or packets of sperm, are introduced into the female's genital aperture during mating.

Once this has been achieved, the male plays no further part in reproduction. The spermatozeugmata, however, behave quite extraordinarily. Some will burst and their sperm will fertilise any ripe eggs that the female may have inside her egg sacs (follicles). Others will become embedded in the ovarian cavity wall, where they will remain until required to fertilise a later batch of eggs, allowing some females to produce numerous broods of fry without the need to mate again.

In these livebearers, the embryos undergo all their development inside the egg sacs – a process known as follicular gestation. Although they are in intimate contact with the female's tissues and possess embryonic structures, such as pericardial sacs which allow for interchange of materials between them, Guppy, Molly and most other Poeciliid fry actually weigh less at birth than the fertilised egg from which they develop.

However, there are approximately 140 species of these Poeciliid fishes (with more being discovered every year) so, hardly surprisingly, variations

Plate 4   A trio of Mosquito Fish (*Gambusia holbrooki*). The top fish is a fully adult male, the middle one is an adult female, and the bottom one is an adolescent male whose anal fin is developing into the characteristic gonopodium found in Poeciliids. (*John Dawes*)

on this basic strategy occur, making strict definitions and groupings virtually impossible.

In fact, all types of embryonic nutrition, from the strict egg-yolk-only variety (lecithotrophy), to pronounced maternal contributions (specialised matrotrophy) are known to occur.

Lecithotrophy is characterised by actual weight loss during embryonic development. Many species exhibit this, e.g. the Guppy, whose new-born fry can weigh as much as 25% less than the eggs from which they develop.

In other species, like some Swordtails (*Xiphophorus*) and Piketop Live-bearers (*Belonesox belizanus*), the fry weigh about the same at birth as the fertilised eggs. Relating this situation to that found in Guppies, it is clear that the potential loss has been made up from somewhere – in this case, maternally derived nourishment. Embryonic growth of this kind is known as 'unspecialised matrotrophic' development.

In the specialised matrotrophs, something rather remarkable occurs. The eggs start off with insufficient yolk to take them through gestation. However, the embryos receive such a plentiful supply of nourishment from their mothers that they can markedly increase their body weight.

In *Poeciliopsis turneri*, for instance, new-born fry are as much as 1,840 times heavier than a fertilised egg. *Heterandria formosa* – one of the Mosquito Fish – goes even further, with its embryos showing a weight increase of around 3,900%! This spectacular achievement is made possible by the production of a pseudo-placenta and by superfoetation – a condition in which a gravid female carries two or more batches of embryos at different stages of development inside her ovaries at one and the same time. A *Heterandria formosa* female, for example, can carry as many as *nine* broods.

The biological advantages of such a strategy are immense. A relatively small female can produce a very large number of offspring over an extended and almost continuous fry-bearing period. Admittedly, the number of fry produced in one go is low, but this is compensated for by the total number of young born over the extended fry-producing period, and by the distinct advantage bestowed on each individual fry in the extremely serious game of survival by its massive weight increase.

It is not just the increase in weight that is important, of course. Growth may perhaps be the most immediately obvious clue that developing fry are being fed by their mother, but there are other advantages that also become apparent when we take a closer look.

For example, fins, musculature, bones and complex structures (such as eyes, mouths and digestive systems) are all able to develop much further as a result of maternal nutrition than would otherwise be the case. It is the sum total of all these advantages that gives these fry such a good start in life, making them highly mobile, keenly-sighted predators from the outset.

In fact, the benefits derived from maternally generated nutrition (plus the almost incidental, but high, degree of protection that internal incubation affords) are so great that fish employing this strategy need to produce only relatively small broods to ensure the survival of the species.

Jenynsia, like the other matrotrophs mentioned above, nourishes its young, but it goes several interesting stages further. One such stage can be seen in the structures, known as trophonemata, that are produced by the females as outgrowths of their ovarian tissues. The trophonemata act almost like a human placenta and confer tremendous advantages on developing *Jenynsia* embryos.

Not all species in which the foetuses develop within the ovarian lumen – as opposed to within the egg sacs, or ovarian follicles – adopt the same approach as *Jenynsia*. In the Goodeids, for example, it's not the mother that develops the feeding (trophic) structures, but the embryos themselves. In this case, the tapeworm- or rosette-like outgrowths (the shape varies from species to species) are known as trophotaeniae, but their overall function is the same as that performed by the trophonemata found in the One-sided Livebearer. The results are also similar – Goodeid fry weighing several thousand times more than the fertilised eggs from which they started developing several weeks earlier.

In the small marine family Parabrotulidae (incorporated by some authorities within the Zoarcidae, the family commonly referred to as Eelpouts), at least one species, *Parabrotula plagiophthalmus*, produces embryonic nutritional structures not unlike the trophotaeniae of the Goodeids – yet, despite this apparent affinity, the fish themselves are as unrelated to each other as one could imagine.

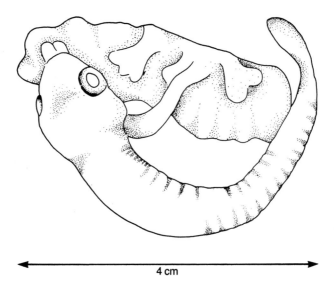

4 cm

Fig. 4   One-sided Livebearer (*Jenynsia lineata*) embryo with trophonemata –
structures developed by the female through which embryos are nourished during
gestation.

The Bythitidae (Viviparous Brotulas) and the closely related Aphyoni-
dae (the former including a few fresh- and brackish-water species among its
total of around 80 or so marine representatives, and the latter comprising
some 18 species, most of which are found at depths exceeding 700 metres)
could not, at first sight, appear to be in any way similar to the shallow-
water, freshwater Goodeids. Yet they, too, contain some species that use
trophotaenial nutrition..

Then, there are the East Russian Comerphoridae (Baikal Oilfishes),
consisting of only two species, both of which are partly lecithotrophic
livebearers; the Embiotocidae (Surfperches), with embryonic weight
increases of up to +24,000%; the Clinidae; and the Labrisomidae, sometimes referred to as Kelpfish or Clinid Blennies, several of which exhibit
superfoetation (as in *Heterandria formosa*, the Mosquito Fish).

This overall picture, involving ovulation, placenta-like structures, huge
increases in body weight and so on, comes a little closer to what most
people would regard as true livebearing – something that is widely associated with mammals but hardly ever with fish.

Attempts at a Definition

Livebearing of one kind or another is known in about 930 species of fish.
Of these, some 420 are Chondrichthyans (cartilaginous species like sharks,
rays, skates and Chimaeras) grouped into 99 genera and 40 families.

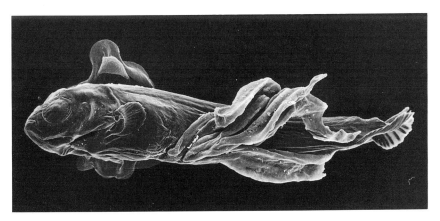

Plate 5 Scanning electron micrograph of a mid-term *Xenoophorus captivus* embryo showing the embryonic trophic structures – known as trophotaeniae – characteristic of Goodeids. (*John Dawes*; reproduced by kind courtesy of the Centre for Electron Optics, University of Bath, Avon, England)

Among the bony fishes (Osteichthys), about 14 families, made up of some 122 genera and 510 species, exhibit livebearing. Then, of course, there is the Coelacanth . . .

It is hardly surprising, therefore, that, when nearly one-twentieth of all known living species of fish produce live young in one way or another, a simple, foolproof definition of livebearing becomes next to impossible.

Even if we were able to define livebearers in an all-encompassing way, we would end up with a definition that would probably not reflect what an aquarist would regard as a livebearer – as opposed to what a scientist would find acceptable. The problem here arises from the differing needs of aquarists and of scientists. To the aquarist, it is important to know what a *livebearer* is. To the scientist, it is probably more important to know what *livebearing* means. So, on one side we have a need to identify *organisms* and on the other we have a need to know and recognise the *process* by which these organisms reproduce.

However, knowing that the problem is difficult should not deter us from trying to find a way round it. Indeed, such attempts are desirable since, in their absence, utter chaos would rule. By trying to come to grips with the issues, we can, at least, bring some order to the chaos, even if we cannot smooth out all the rough edges.

In fact, many eminent people have tried, with varying degrees of success. Two ichthyologists whose work in this direction I find particularly interesting are Eugene Balon and Miles Keenleyside. Each has attempted to make sense of the mind-shattering array of reproductive strategies that exist in fish by trying to break the spectrum down into workable portions. In so doing, they have provided a useful framework that can be used to incorporate virtually all the permutations imaginable. Keenleyside's categories and Balon's 'Ecoethological Guilds' can be matched up to a certain extent, the amalgam producing quite a manageable summary of reproductive techniques (see Table I below).

As can be seen from Table I, livebearers would come under Keenleyside's Category 2, 'Mating with internal fertilisation' and Balon's Guild 3b, 'Internal bearers'.

TABLE I   COMPARISON BETWEEN KEENLEYSIDE'S CLASSIFICATION AND BALON'S ECOETHOLOGICAL GUILDS

KEENLEYSIDE	BALON

1. Mating with external fertilisation:
   a) Species with pelagic (free-drifting) eggs
   b) Species with demersal (bottom or substratum) eggs – no prolonged guarding
   c) Species with demersal eggs and prolonged guarding
   d) Species that carry their eggs
2. Mating with internal fertilisation
	1. Non-guarders:
	   a) Open substratum spawners
	   b) Brood hiders
	
	2. Guarders:
	   a) Substratum choosers
	   b) Nest spawners
	
	3. Bearers:
	   a) External bearers
	   b) Internal bearers

Derived from: Balon (1975) and Keenleyside (1979).

One thing that will become immediately apparent from Table I is that livebearing accounts for quite a small part of a much greater whole. This, though, should not come as any great surprise, however devoted we may be to livebearers. As I mentioned earlier, livebearing occurs in slightly fewer than one in twenty species of fish. Furthermore, diverse though the varying degrees of livebearing outlined in the foregoing sections of this chapter may be, they are nowhere near as diverse as the range of egglaying strategies that exist.

Balon subdivided his 'Internal bearers' in accordance with their reproductive styles:

  i) ovi-ovoviviparity – internal fertilisation with subsequent deposition of fertilised eggs
 ii) ovoviviparity – internal fertilisation and retention of eggs up to parturition, with all the embryonic nourishment being derived from yolk
iii) viviparity – internal fertilisation and retention of eggs with embryonic nutrition being supplied partly or wholly by the mother

Now, at last, it seems as if we are getting closer to being able to formulate a definition of a livebearer. This would indeed be so, were it not for a very important point that has repeatedly been highlighted by a number of ichthyologists. The fact is that it is impossible to separate ovoviviparity and viviparity as implied in the foregoing categories. As John Wourms pointed out in his classic review of viviparity in 1981, 'In many instances the maternal–foetal metabolic relationship is unknown, hence no distinction can be made.' Further, the massive body of data accumulated over the years regarding reproduction in the cartilaginous fish and in the bony fish, shows that we are dealing with a continuum that starts at no dependence whatsoever on maternally generated nutrition (and total dependence on yolk) and ends at almost total dependence on maternal contributions (and none, or very little, on yolk).

In the end, John Wourms came to the conclusion that 'the only operational distinction is that between oviparity and viviparity'. He went on to say: 'Viviparity should now be defined as a process in which eggs are fertilised internally and are retained within the maternal reproductive system for a significant period of time, during which they develop to an advanced state and are then released.'

There are areas of contention even within such a wide-ranging definition, of course. For example, how does one define 'for a significant period of time'? Or, how does 'an advanced state' differ from a non-advanced one?

Even so, some welcome degree of simplification is brought to the chaos by, at least, one major feature of this definition. To quote Wourms once more: 'Since the source of embryonic nutrients is not considered, this revised definition now includes the terms "ovoviviparity" and "viviparity" as used in the older sense.'

In fact, it goes even further than this, because those fish which appear under Balon's ovi-ovoviviparous category, e.g. *Oryzias*, *Tomeurus*, *Horaichthys*, etc., now become regarded as exhibiting 'facultative viviparity', with all the others considered to exhibit 'obligate viviparity'.

Therefore, from the scientific point of view, we now seem to have a workable definition of viviparity which does not lay down as a prerequisite the internal retention of fertilised eggs up to the moment of birth. As long as eggs are fertilised internally and begin their development within the body of the female, this is sufficiently significant to qualify as some form of viviparity.

One of the beauties of this definition is that it cuts across genera and families. We are, consequently, no longer bound by this sometimes restrictive parameter.

From the aquarists' point of view, however, the above definition, workable though it may be scientifically, still leaves us with a few problems.

One reason, as mentioned earlier, is that aquarists need to know what a livebearer actually is. To say that a livebearer is a fish that exhibits viviparity – be it facultative or obligate – doesn't help a great deal, since this would imply that the hobby of keeping livebearers would have to take into account fish as different from each other as a Guppy, a Blue Shark, a Livebearing Blenny and, even the Coelacanth itself – clearly an impossible task.

We therefore need to narrow the definition, restricting its boundaries in the first instance to species that are kept in aquaria. Inevitably, we hit several snags in adopting this approach (including just what constitutes an

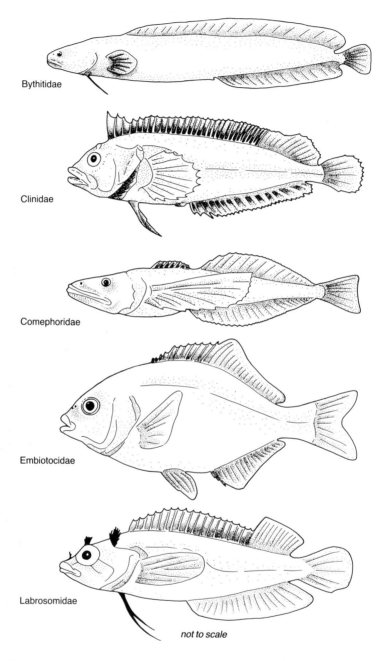

Fig. 5 Livebearing of one degree or other occurs in a number of families of bony fishes, not all of which are closely related to each other in evolutionary terms. This diagram shows livebearing representatives of five distinct families. (Based on J. Nelson, *Fishes of the World*, 2nd ed., John Wiley 1984)

'aquarium' species), but they are not insurmountable.

Cutting down the definition to aquarium species eliminates the problems of the Coelacanth, the Oilfish, Viviparous Brotulas, Parabrotulids, Surfperches, Kelpfish, *Horaichthys*, most sharks, skates and rays, and the Chimaeras. It does not, however, eliminate some other sharks and rays, or Scorpionfish or the Eelpouts. Of these, the Scorpionfish are undoubtedly the ones kept most frequently by hobbyists, but the other two are also kept, admittedly much less frequently. In addition, with more and more species being kept by hobbyists nowadays, the above list is likely to change with time, perhaps, for example, incorporating *Horaichthys* in due course.

Diverse though this group may be, it is nevertheless considerably more manageable than the former, all-embracing one. It becomes even more manageable if we exclude all the marine fish as the next step. In aquarium terms, this is quite in order because enthusiasts tend to see themselves in the first instance, as marine or freshwater hobbyists, irrespective of what their specialisation within one or the other of these broad fields may be. It is once within these two categories that fine refinements occur, such as invertebrate/fish-only/mixed community aquarists on the marine side, or coldwater/tropical on the freshwater side, with further subdivisions into, say, Catfish, Cichlid, Killifish, Koi, Goldfish, Livebearer keepers or whatever, further down the line.

Having eliminated all the non-aquarium and marine species, we are now left with a motley collection of freshwater ones. Some appear to fall fairly and squarely within the widely-accepted view of what a livebearer is, e.g. Anablepids, Goodeids, Poeciliids and Hemiramphids, but others don't fit in too comfortably, in particular, the Swordtail Characin (*Corynopoma*), the two Killies mentioned earlier and the Medaka.

At this point, an important decision needs to be taken.

Traditionally, the term livebearer has not included fish like the Swordtail Characin, the two 'abnormal' Killies or the Medaka. It has, however, included the aberrant Poeciliid *Tomeurus gracilis*, generally regarding it as merely an acceptable exception to the rule.

In my mind, there doesn't seem to be too great a difference between the facultative livebearing habits of *Tomeurus* and those of the Killifish *Cynolebias brucei* and *C. melanotaenia*. They are even pretty closely related, *Tomeurus* belonging to the so-called Livebearing Toothcarps (the Livebearers) and the two others to the Egglaying Toothcarps (the Killifishes).

If we accept this similarity, then widening the boundaries of our potential definition to include the two Killies is only a matter of degree.

Taking things a significant step further, there would then appear to be a case for assimilating even the Swordtail Characin and the Medaka, which, after all, also exhibit facultative viviparity.

Some would undoubtedly find such a move totally unacceptable. But others would argue that, having trimmed down the breadth of the aquarium definition of a livebearer to just freshwater aquarium species that exhibit facultative or obligate viviparity, we cannot now start imposing new restrictions by excluding a fish simply on the grounds that it is a Characin or a Medaka.

A further twist, which adds considerable strength to this viewpoint, comes from the classification of Cyprinodontiform fishes (which include the Killies, Anablepids, Poeciliids and Goodeids) proposed in 1981 by Lynne Parenti. In this classification, not all the fish within the families Poeciliidae, Goodeidae or Anablepidae are livebearers. Therefore, if not even all the members of these families are livebearers any longer, how can we justify the exclusion of *Corynopoma riisei* and *Oryzias latipes* from any list of livebearers covered by the terms of reference of any definition, and still include *Tomeurus*? My personal inclination is towards including all of them, despite the fact that some established livebearer keepers would probably not consider these inclusions as true members of the livebearer club.

Tying up all the loose ends discussed in the foregoing pages, it is now possible to put together some form of working definition of a livebearer

Plate 6 Mollies are usually sold as suitable community fish. However, Sailfin varieties in particular, require brackish water conditions for long-term health and survival. (*Harry Grier/Florida Tropical Fish Farms Association*)

that most aquarists should be able to apply in practice. One relatively technical version of the definition would be:

A livebearer is a freshwater aquarium fish that exhibits facultative or obligate viviparity, as defined by Wourms (1981).

A less technical, but slightly longer, version of this definition would run something like:

A livebearer is a freshwater aquarium fish that employs internal fertilisation of eggs followed by either their deposition after a shorter or longer period of time or their retention within the body of the female until the moment of birth.

I cannot pretend for a moment that these are foolproof statements – not when obvious livebearers such as the Four-eyed Fishes (*Anableps* spp) and the Mosquito Fish (*Gambusia affinis*) can exist quite happily in brackish water in the wild, or when Sailfin Mollies (*Poecilia latipinna* and *P. velifera*) can even occur in coastal waters. Nevertheless, these species can, and do, live in freshwater as well and are, in any case, generally regarded as freshwater species with special requirements.

Further Reading

Balon, Eugene K., 'Reproductive guilds of fishes: a proposal and definition', *J. Fish. Res. Board Can.* 32, pp. 821–64 (1975).

Balon, Eugene K., 'Patterns in the evolution of reproductive styles in fishes' in *Fish Reproduction – Strategies and Tactics*, G. W. Potts and R. J. Wootton (eds.), Academic Press p. 410 (1984).

Keenleyside, Miles H. A., 'Breeding Behaviour', in *Diversity and Adaptation in Fish Behaviour*, Springer-Verlag p. 208 (1979).

Meyer, Manfred K., Wischnath, Lothar and Foerster, Wolfgang, *Lebendgebärende Zierfische – Arten der Welt*, Mergus-Verlag p. 496 (1985).

Wourms, John P., 'Viviparity: the maternal–fetal relationship in fishes', *Amer. Zool.*, 21, pp. 473–515 (1981).

Classification

Most books on aquarium livebearers restrict their coverage to species belonging to five traditional families: Poeciliidae (Poeciliids), Goodeidae (Goodeids or Mexican Livebearers), Hemiramphidae (Halfbeaks), Anablepidae (Four-eyed Fishes) and Jenynsiidae (One-sided Livebearers).

For reasons that will, hopefully, become clearer as this chapter unfolds, the above will now be considered as being represented by *parts* of four, instead of five, families in the present work, these families being the Poeciliidae, Goodeidae, Hemiramphidae and Anablepidae (incorporating the Jenynsiidae). There will also be the following additions: *Corynopoma riisei* (family Characidae), *Oryzias latipes* (family Oryziidae), and *Cynolebias brucei* and *C. melanotaenia*, (traditionally classed as Cyprinodontidae but now regarded as belonging to the family Rivulidae).

All these, with the exception of *Corynopoma* and *Oryzias*, came under close scrutiny (not for the first time, of course) in 1981, when Lynne Parenti, working in the Department of Ichthyology at the American Museum of Natural History, undertook a detailed study of the Cyprinodontiform fishes as part of her Doctor of Philosophy research programme within the Biology Faculty of the City University of New York. Her findings, conclusions and proposals more or less turned livebearer and killifish classification on its head, shaking out old beliefs and bringing in revolutionary new ones which, to be fair, are not yet universally accepted.

To appreciate the implications of Parenti's classification – which I have adopted in this book – a brief examination of the old, superseded one, may not come amiss.

First, though, I will deal with the two families that contain the most controversial of the species I have brought under the overall livebearing umbrella.

Characidae (Characins)

This is a very large family of fish regarded by many authorities, including Joseph Nelson, as containing around 166 genera and 840 species. These are further subdivided into a number of subfamilies, one of which (the largest) is the Characinae, to which about 142 of the 166 genera belong. Among these are the Tetras, such as the Neon and Cardinal (*Paracheirodon innesi* and *P. axelrodi*, respectively), the famous Blind Cave Tetra (*Astyanax fasciatus mexicanus*, formerly *Anoptichthys jordani*) and the Swordtail Characin (*Corynopoma riisei*).

Plate 7   The Swordtail Characin (*Corynopoma riisei*) is the only member of its family that exhibits internal fertilisation. The specimen illustrated is a mature male. (*Bill Tomey*)

Other ichthyologists, notably Jacques Géry, consider Nelson's subfamily Characinae as a family in its own right, subsequently subdividing it into a new complement of subfamilies.

One of these is the subfamily Glandulocaudinae which, even according to Géry, 'is probably not [a] natural [group], but . . . an artificial assemblage of, for practical purposes, a series of forms coming from different stems'. Not surprisingly, perhaps, this heterogeneity is reflected in the further subdivision of the subfamily into two tribes, the Glandulocaudini (the Croaking Tetras – so called because some species can produce such sounds) and the Xenurobryconini (the Bristly-mouthed Tetras – so called because of the numerous, minute conical teeth that give the fish their very characteristic mouths).

According to Géry's classification, *Corynopoma riisei* belongs to the Glandulocaudinae, which are distinguished from their closest relatives by having moderate body size, terminal or superior mouths, and heavy teeth – usually tri- to penta-cuspid (having three to five cusps), arranged biserially (in two rows) on the upper jaw.

Characoid classification, at least at the family and subfamily level, is currently, as it has been for a long time, in a state of flux. Both the main versions will therefore be quoted in the entry for *Corynopoma* in Part III.

For comparison's sake, though, Table II shows how this species fits into both Nelson's and Géry's classifications.

TABLE II   *CORYNOPOMA RIISEI* CLASSIFICATION

NELSON

Family   : Characidae
Subfamilies: Alestiinae   (African Tetras)
                Crenuchinae   (Northern South American Tetras)
                Serrasalminae (Piranhas and relatives)
                Characinae   (South American Tetras)
                   Genus: *Corynopoma*
                      Species: *riisei*

GÉRY

Family   : Characidae (=Nelson's Characinae)
***Subfamilies:** Agoniatinae
                Rhaphiodontinae
                Characinae
                Bryconinae
                Clupeacharacinae
                Paragoniatinae
                Aphyocharacinae
                Stethaprioninae
                Tetragonopterinae
                Rhoadsiinae
                Glandulocaudinae
                   Tribes: Xenurobryconini (Bristly-mouthed Tetras)
                            Glandulocaudini (Croaking Tetras)
                      Genus: *Corynopoma*
                      Species: *riisei*

* All the subfamilies are American in their distribution.

Oryziidae (Medakas or Ricefishes)

Variously referred to as the Oryziatidae or Oryziidae, the Medakas are fresh- and brackish-water fish distributed from India and Japan to the Indo-Australian Archipelago.

According to Nelson (1984), the Medakas represent one of three families that make up the suborder Adrianichthyoidei, which numbers only four genera and eleven species.

The two other families are the Adrianichthyidae and, very interestingly, the monotypic family (one genus, one species) Horaichthyidae. *Horaichthys setnai*, as discussed in the previous chapter, shares the characteristic of facultative viviparity (internal fertilisation with subsequent egg deposition) with *Oryzias latipes*.

Although there are distinct differences between the three families as designated in Nelson (1984), e.g. 'Jaws not tremendously enlarged' in the Oryziidae, 'Jaws tremendously enlarged' in the Adrianichthyidae, and a highly complex copulatory organ in the Horaichthyidae, there are pronounced similarities as well. None, for instance, possess a lateral line organ on the body. Probably more significantly, none possess the following skull bones: the supracleithrum, metapterygoid or ectopterygoid. In addition, none possess a protrusible (protractible) upper jaw.

These similarities were felt to be so significant by Donn Rosen and Lynne Parenti that, in 1981, they suggested that all three families be amalgamated into one, the Adrianichthyidae, although previously (1966) Rosen, working with Humphry Greenwood, Stanley Weitzman and George Myers, had classified all three separately, just as Nelson did in 1984.

As in the case of the two classifications of Characins in the previous section, both approaches in the classifications of *Oryzias latipes* appear to have genuine strengths. The entry for *Oryzias* in Part III will therefore, and for the same reasons as in *Corynopoma*, recognise both.

Cyprinodontiforms (Killifishes and Livebearers)

Traditionally, the Cyprinodontiform fish have been referred to as Toothcarps and divided into two categories – the Egglaying Toothcarps, or Killifishes, and the Livebearing Toothcarps, or Livebearers.

As I hope to have demonstrated in the previous chapter, these rigid subdivisions have been shown to be only partly applicable. When such a situation exists, the actual *number* of exceptions to the rule becomes irrelevant. It is their *significance* that is important. Again, I hope to have shown that these exceptions are very significant indeed – sufficiently so for numerous workers in this field of study to question the validity of the old classifications and, subsequently, propose alternatives.

A major landmark in the classification of livebearers was established in 1963 with the publication of Donn Rosen's and Reeve Bailey's detailed review of Poeciliids and their relationships.

They ended up by dividing the family Poeciliidae into three subfamilies: the Tomeurinae, containing a single species, *Tomeurus gracilis*; the Poeciliinae, comprising 19 genera grouped into 5 tribes; and the monotypic Xenodexiinae, represented by *Xenodexia ctenolepis*.

One of the most revolutionary and significant outcomes of the study was that it brought together the widely scattered, and often conflicting, Poeciliid work of previous ichthyologists into a more rational and useable format. This resulted in a substantial reduction in the total number of genera through the synonymising of hitherto differently named, but biologically indistinguishable, fish previously believed to belong, or actually described as belonging, to separate species or genera.

Such was the influence of Rosen's and Bailey's work that it stood, without being challenged in any major way, for about eighteen years. Not

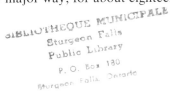

till 1981, when Lynne Parenti published her analysis, was it ever remotely likely to undergo fundamental revision. In fact, it would be true to say that, despite the thoroughness and overall soundness of Parenti's work, Rosen's and Bailey's classification of Poeciliids is still adhered to by many hobbyists, scientists and writers of aquarium literature. I, though, find Parenti's arguments very convincing indeed and will adopt her classification throughout this book.

The Cyprinidontiforms are thought by Parenti to have a single line of descent, i.e. they are thought to be monophyletic. This is believed to be reflected in the skeletal characters they share, some of which are quite complex and are therefore unlikely to have arisen totally independently in all these fish if they had had several evolutionary lines of descent rather than a single one.

Aspects of skull, gill, jaw, pelvic girdle, tail and rib skeletal structures all show shared characteristics that point strongly towards some pronounced degree of relatedness. Added to these physical characteristics there is the more physiological one of an extended period of embryonic development, be that in externally laid eggs (as in the vast majority of Killifish) or in internally retained ones (as in all the livebearers except one).

Applying the same criterion of shared characteristics, Parenti concluded that the viviparous families were all monophyletic. However, the family hitherto known as the Cyprinodontids (Cyprinodontidae) and containing all the Killifishes, has dual ancestry. This means that it needs to be split in a way that reflects this.

The problem with splitting is that you can sometimes create complications that hinder rather than help. For example, the traditional family Cyprinodontidae contained eight subfamilies. If, as Parenti concluded, we need to split the Cyprinodontids into two, these subdivisions would themselves end up as subfamilies. What happens to the previous eight subfamilies? Do they become tribes? And, if so, does this reflect their real significance?

One solution – the one adopted by Parenti in this case – is to *raise* the status of some of the units, and create a new usage for them. As a result, Parenti's split shows the order Cyprinodontiformes as consisting of two suborders, the Aplocheiloidei and the Cyprinodontoidei, the latter containing fewer representatives than in the past (these having been transferred to the Aplocheiloidei).

This simple move solves a lot of problems because it opens up a way of accommodating all the Killies, plus the livebearers, further down the line.

As we move down this line, things start getting very interesting indeed because we begin to find proposals that challenge our concepts of how the livebearers that we have become so familiar with over the years fit into the larger scheme of things. Perhaps the greatest challenge of all is to the importance that we've traditionally given to livebearing itself.

This type of reproduction has always been regarded as a key factor separating the Livebearing Toothcarps from their egglaying counterparts. But to quote Parenti, 'In the present study, this presumption was discarded at the outset'. Her several reasons for this include:

a) internal fertilisation *per se* is not necessarily indicative of a close biological relationship between species;
b) egg retention up to the moment of birth is similarly independent;
c) special fin modifications employed in internal fertilisation have arisen many times in totally unrelated groups;
d) follicular gestation occurs independently in some Poeciliids, and in *Anableps*, the Four-eyed Fish.

If one accepts the validity of these arguments, then it is difficult not to accept Parenti's overall conclusions.

Therefore, if internal fertilisation, the presence of follicular gestation, embryonic structures, etc., are not, individually, as important as previously thought, then, by the time we feed into the mix other factors such as the skeletal similarities, a new, and very interesting picture of the classification of these fishes begins to emerge.

As a good example of this, we could take the Four-eyed Fish (*Anableps*) and the One-sided Livebearer (*Jenynsia*). Traditionally, these two genera have been classified as separate families, the Anablepidae and Jenynsiidae respectively.

At the top end of Parenti's classification, the family Anablepidae is distinguished from all the other Cyprinodontiform families by very distinctive bone characteristics around the eyes and by very specific dental arrangements, particularly in juveniles or embryos.

Such a list of distinguishing features brings both *Anableps* and *Jenynsia* into the same family. Interestingly, when we add one other unusual criterion – left or righthandedness – we find that a hitherto unconsidered fish, *Oxyzygonectes*, is also brought into the fold.

Plate 8   A Guppy female giving birth. Livebearing, according to Lynne Parenti's argument, is not the key factor in fish classification that it has traditionally been accepted to be. (*Harry Grier*)

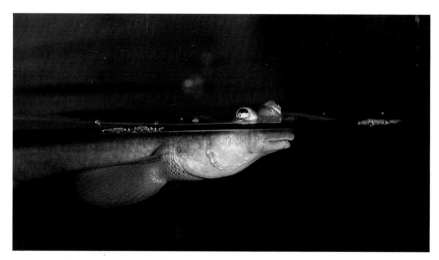

Plate 9   The Four-eyed Fish (*Anableps anableps*) now shares its family and subfamily with the One-sided Livebearer (*Jenynsia lineata*). (*Bill Tomey*)

Unlike *Anableps* and *Jenynsia*, *Oxyzygonectes* is an egglayer, but one in which the males have 'a distinct anal papilla which, in preserved specimens, has been offset to the left or the right'.

Even if this were something that occurred in these specimens as an accident of preservation, *Oxyzygonectes* females have 'a pocket of scales' surrounding the anus and the bases of the first few anal fin rays – a feature which is thought to be characteristic of the Anablepids.

Parenti's interpretation of these features (which I have only summarised here) led her to conclude that *Oxyzygonectes* was more closely related to *Anableps* and *Jenynsia* than earlier classifications, which placed it within the Killifish subfamily Fundulinae, indicated.

However, and despite the lower significance given to viviparity by Parenti, the fact that *Oxyzygonectes* is a typical egglayer, and not a livebearer as its nearest relatives are, shows that the relationship, while close, is not sufficiently so to warrant total inclusion. The logical step, therefore, is to set up two separate subfamilies within the Anablepidae. One, the Anablepinae, now incorporates *Anableps* and *Jenynsia*, the two livebearers, while *Oxyzygonectes* is given a family all to itself, the Oxyzygonectinae (see Table III opposite).

It is clearly impossible to go through all the analyses, comparisons and individual conclusions which, together, constitute Parenti's major revision of the Cyprinodontiforms. For full details, I would urge the interested reader to obtain a copy of the original work (see Further Reading at the end of this chapter). It is heavy going but, in my opinion, well worth the effort, and a must for anyone who would like to specialise in livebearers.

Parenti's analysis takes the reader step-by-methodical-step through the complicated field of Cyprinodontiform systematics ending, after more than

200 pages, with a summary classification of the group, an abridged version of which appears in Table IV.

Taking the relevant families and subfamilies in turn at this stage should help round off the Cyprinodontiform classification picture.

TABLE III   SUMMARY OF THE ANABLEPIDAE (ACCORDING TO PARENTI, 1981)

Family	Anablepidae
Subfamily	Anablepinae (incorporating the former families Anablepidae and Jenynsiidae)
Genera	*Anableps*, *Jenynsia*
Subfamily	Oxyzygonectinae (new subfamily)
Genus	*Oxyzygonectes*

TABLE IV   ABRIDGED CLASSIFICATION OF CYPRINODONTIFORMS (ACCORDING TO PARENTI, 1981)

Order	Cyprinodontiformes
Suborder	Aplocheiloidei
Families	Aplocheilidae
	Rivulidae*
Suborder	Cyprinodontoidei
Families	Profundulidae
	Fundulidae
	Valenciidae
Superfamily	Poecilioidea
Family	Anablepidae*
Subfamilies	Anablepinae*
	Oxyzygonectinae
Family	Poeciliidae*
Subfamilies	Poeciliinae*
	Fluviphylacinae
	Aplocheilichthyinae
Superfamily	Cyprinodontoidea
Family	Goodeidae*
Subfamilies	Empetrichthyinae
	Goodeinae*
Family	Cyprinodontidae
Subfamilies	Cubanichthyinae
	Cyprinodontinae

* The Cyprinodontiform livebearers featured in this book all belong to these families and subfamilies.

Rivulidae

The inclusion of this family in a book on livebearers may not find universal agreement among either aquarists or ichthyologists. However, having adopted John Wourms' definition of viviparity earlier, the existence of two species of Killifish that exhibit facultative viviparity within this family warrants its inclusion here.

The two species in question, *Cynolebias brucei* and *C. melanotaenia*, have traditionally been classified as members of the subfamily Rivulinae. In her revision, Parenti elevated the group to the rank of family, giving it equal status to her other family, the Aplocheilidae, which includes such well-known genera as *Nothobranchius*, *Pachypanchax*, *Aphyosemion* and *Epiplatys*, some of the classic Killies of the aquarium hobby.

The Rivulidae, for their part, can also boast a few famous genera within their ranks, notably *Rivulus* and *Cynolebias*.

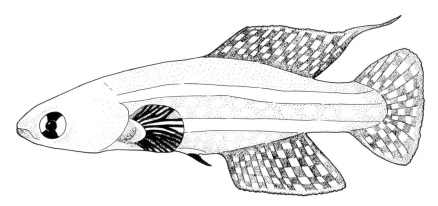

Fig. 6   *Cynolebias melanotaenia* – one of two Killifish species that exhibit facultative viviparity.

TABLE V   SUMMARY OF THE APLOCHEILOIDEI (ACCORDING TO PARENTI, 1981)

Suborder	Aplocheiloidei (new usage for the term)
Family	Aplocheilidae
Genera	*Adamas, Aphyosemion, Aplocheilus, Epiplatys, Fundulopanchax, Nothobranchius, Pachypanchax*
Family	Rivulidae
Genera	*Austrofundulus, Cynolebias*, Neofundulus, 'Neofundulus', Pterolebias, Rachovia, Rivulus**, 'Rivulus', Trigonectes*

 * The two species considered in detail in this book belong to this genus.

** Contains the self-fertilising hermaphroditic species *R. marmoratus*.

A combination of skeletal features, some shared, some unique and one – the postcleithrum, one of the bones of the pectoral girdle – totally lacking, separates the Rivulidae from all the other Aplocheiloids.

The genus *Cynolebias* contains about 35 species, of which all but two are typical egglayers.

Anablepidae

Details of the internal re-structuring of this family were discussed earlier and will, therefore, not be repeated here.

Of the two subfamilies, the Anablepinae and the Oxyzygonectinae, the former is livebearing, while the latter is egglaying.

The livebearing Anablepids exhibit two very different methods of embryonic nutrition. *Anableps* employs intra-follicular gestation, i.e. embryos develop within ovarian follicles or egg sacs embedded in the ovarian wall. In *Jenynsia*, the only other genus in the subfamily, females ovulate, with the embryos subsequently developing within the ovarian cavity and being nourished through maternally produced structures called trophonemata which act as a pseudo-placenta.

Despite these differences, though, both genera are distinguished from all other Cyprinodontiforms 'by having thickened and elongated anal rays in males which are twisted around each other and covered by a fleshy tube, tubular sperm duct [and] gonopodium offset either to the left or the right' (Parenti, 1981).

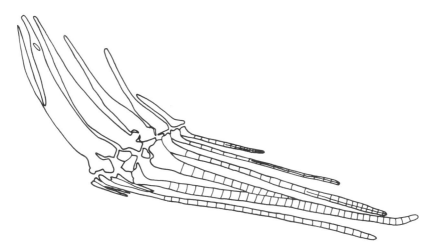

Fig. 7   Anablepids are distinguished from other livebearers by their very distinctive gonopodia. This is the gonopodium of the One-sided Livebearer (*Jenynsia lineata*). (After L. Parenti, *Bull. Am. Mus. Nat. Hist.*, Vol. 168, 1981)

Poeciliidae

Following Parenti's 1981 re-organisation, the family Poeciliidae now consists of three subfamilies: the viviparous Poeciliinae (with one exception, *Tomeurus gracilis*) and the oviparous Fluviphylacinae and Aplocheilichthyinae (Lamp-eyes).

Further work by Lynne Parenti, this time in collaboration with Mary Rauchenberger, led, in 1989, to a refinement of the classification of the Poeciliinae – something that had remained largely unaltered since Rosen's and Bailey's review of 1963.

Parenti and Rauchenberger based their re-classification, as far as possible, on relationships within the group. One result of adopting this approach was that the two major types of viviparity found within the Poeciliines had to be given due recognition. Therefore, *Tomeurus gracilis* had to be separated from the others in some way.

As the Poeciliinae constitute a subfamily, it might be thought logical to create two tribes, one to accommodate *Tomeurus* and the other for the remaining genera and species. But the moment this decision is taken new problems arise. For instance, within the main group of Poeciliines, various subgroups can be identified, such as one incorporating the Mollies, Limias and Guppy, another embracing the somewhat similar Gambusias, the *Brachyraphis* species and their immediate relatives, and so on. Taking this line of thinking one significant step further, the mere fact that Mollies, Limias, the Guppy and their closest relatives can, together, be seen as a subgroup means, conversely, that the subgroup itself can be subdivided into at least three sub-subgroups of species. The same argument can be applied to nearly all the Poeciliine subgroups, of course.

As with the Rivulines discussed in an earlier part of this chapter, such a situation can be resolved quite neatly by re-evaluating the status of the various groups and subgroups.

Plate 10   Several subgroups can be distinguished among the Poeciliids, including the various *Brachyraphis* species. This is *B. rhabdophora*. (*Dennis Barrett*)

TABLE VI   CLASSIFICATION OF POECILIINE FISHES (ACCORDING TO PARENTI AND RAUCHENBERGER, 1989)

Subfamily	Poeciliinae (=family Poeciliidae of Rosen & Bailey, 1963)
Supertribe	Tomeurini
Genus	*Tomeurus*
Supertribe	Poeciliini
Tribe	Poeciliini
Genus	*Alfaro*
Genus	*Poecilia*
Subgenera	*Lebistes, Limia, Odontolimia**, Pamphorichthys, Poecilia*
Genus	*Priapella*
Genus	*Xiphophorus*
Tribe	Cnesterodontini
Genus	*Cnesterodon*
Genus	*Phalloceros*
Genus	*Phalloptychus*
Genus	*Phallotorynus*
Tribe*	Scolichthyini
Genus*	*Scolichthys*
Tribe	Gambusini
Genus	*Belonesox*
Genus	*Brachyraphis*
Genus	*Gambusia*
Subgenera	*Arthrophallus, Gambusia, Heterophallina*
Tribe	Girardinini
Genus	*Carlhubbsia, Girardinus, Quintana*
Tribe	Heterandrini
Genus	*Heterandria*
Subgenera	*Heterandria, Pseudoxiphophorus*
Genus	*Neoheterandria*
Genus	*Xenophallus*
Genus	*Poeciliopsis*
Subgenera	*Aulophallus, Poeciliopsis*
Genus	*Priapichthys*
Genus	*Pseudopoecilia*
Genus	*Phallichthys*
Tribe	Xenodexiini
Genus	*Xenodexia*

Based on Parenti and Rauchenberger in Meffe/Snelson: *Ecology and Evolution of Livebearing Fishes* (1989).

  * The tribe Scolichthyini and genus *Scolichthys*, erected by Rosen in 1967, are the only new tribe and genus named since Rosen's and Bailey's revision of 1963.

** *Odontolimia*, erected by Rivas in 1980 as a subgenus of *Limia* (itself regarded as a subgenus by Parenti and Rauchenberger) is the only new subgenus to be named since Rosen and Bailey's 1963 review of Poeciliid fishes.

Parenti and Rauchenberger reappraised the Poeciliines in this light and came up with a re-classification of the whole group (see Table VI on page 41) that probably reflects the relationships between its 190 or so members somewhat more realistically than before. Their new classification has yet to receive widespread acceptance and they themselves point out that it 'is in no way meant to substitute for a needed comprehensive systematic revision and biographic analysis of Poeciliine fishes'. But it provides a working framework within which relations between species, genera and other identifiable subgroups can be fitted in in a way that makes considerable sense.

Goodeidae

Members of this family have, traditionally, been referred to as Goodeids or, by hobbyists, as the Goodeas – a name that should be discouraged since it seems to imply that all Goodeids belong to the genus *Goodea*. Two other names occasionally encountered in the literature are Mexican Livebearer or, more rarely, Highland Carp. Again, both these terms should be avoided. The first is inaccurate because many livebearers, in addition to the Goodeids, are found in Mexico and this label therefore tells us next to nothing of significance. The second is misleading because Goodeids are predominantly but not exclusively 'highland' in distribution (*Ilyodon*, for instance, is a lowland genus) and because the term 'carp' might lead to some confusion, since it is normally used as the common name for *Cyprinus carpio*. (Even if, as is actually the case, the term carp is used in recognition that Goodeids are Toothcarps, there should be some indication to this effect, with the further qualification that we are referring to Livebearing Toothcarps.)

All in all, it would seem more appropriate and more accurate to refer to these interesting fish simply as Goodeids.

This term needs some qualification, though. In the past, the only fish that would have qualified as Goodeids would have been the 35–40 livebearing species that belong to the 17 genera that can be distinguished from Poeciliines and other livebearers by the possession of a notched anal fin in the males.

Following Parenti's 1988 re-classification of Cyprinidontiform fishes, however, the family Goodeidae was expanded to include two egglaying genera from the Death Valley system and eastern Nevada, *Empetrichthys* and *Crenichthys*. But there are other features that separate these two genera from the rest of the Goodeidae. They both, for example, lack pelvic fins and pelvic fin supports.

What we have, as a result, is a situation rather like that encountered within the Anablepids and with a similar solution. Parenti divided the family Goodeidae into two subfamilies – the Empetrichtyinae, containing *Empetrichthys* and *Crenichthys*, and the Goodeinae, containing the remaining genera. The egglaying Empetrichtyinae will receive no further consideration in this book. It is only the Goodeinae – which, for the sake of simplicity, will still be referred to as Goodeids – that interest us.

Plate 11   Goodeids (this is *Characodon lateralis*) are sometimes, confusingly, referred to as Mexican Livebearers or Highland Carp. Both names should probably be discontinued. (*'Aquarian' Fish Foods*)

These remarkable fish were virtually unknown within the aquarium hobby until Robert Rush Miller and John Michael Fitzsimons (1971) described *Ameca splendens*, the species that really introduced the Goodeids to aquarists. This was so different from other livebearers that there was an immediate explosion of interest among livebearer hobbyists, both in Europe and the United States, resulting in the appearance of many hitherto unseen species within the space of a few years. Today, interest in Goodeids continues to expand, but the feverish activity of the mid-to-late 1970s has died down to a more manageable level.

Goodeid males, as already mentioned, have a notched anal fin instead of a gonopodium, but this is just the most obvious of many significant differences between these and other livebearers.

Perhaps the most distinctive feature of the Goodeids is their method of embryonic nutrition. After an egg has been fertilised within its ovarian follicle, it is ejected into the ovarian lumen, where the embryo completes its development bathed in the surrounding ovarian fluid. This fluid is rich in nutrients and is therefore, potentially, an invaluable source of food for the embryos. The problem is how adequately to exploit this food source. Ingestion through the gill openings is an obvious possibility, but structures like the trophonemata found in the One-sided Livebearer (*Jenynsia lineata*) seem a far better option. In Goodeids, the solution is provided by structures known as trophotaeniae, which are produced from the vents of the embryos themselves, instead of from the ovarian wall of the female as in *Jenynsia*.

So there are at least two criteria that separate Goodeids from all other livebearers – the possession of a notched anal fin in males and the feeding of embryos by means of trophotaeniae. Up to quite recently, we had an exception to the 'trophotaenial rule', more or less in the same way as *Tomeurus gracilis* is regarded as an exception to the fundamental rule of internal egg retention in Poeciliines. The exception was *Ataeniobius toweri*, which, as its generic name implies, was regarded as not possessing trophotaeniae. This lack, in a fish that exhibits so many of the other characteristics of the subfamily, has been repeatedly considered sufficiently significant by workers in the field of livebearer classification to warrant placing *Ataeniobius* in a subfamily of its own, the Ataeniobiinae.

Taking ovarian and trophotaenial characteristics together, the remaining Goodeids have traditionally been allocated to three other subfamilies:

Characodontinae	containing the sole genus *Characodon*
Girardinichthyinae	containing genera such as *Girardinichthys*, *Skiffia*, *Ilyodon*, *Neotoca* and others
Goodeinae	containing genera such as *Allotoca*, *Ameca*, *Chapalichthys*, *Goodea*, *Xenoophorus*, *Xenotoca* and others

This classification, logical though it seems, can be applied only if the Goodeids constitute a family. So, when Parenti lowered the status of the livebearing Goodeids to subfamilial level, it was no longer valid. At the very least the subfamilies would have to be brought down to tribe level, or an equivalent (as in the Poeciliines).

Ataeniobius, as we have seen, is a further complication, as the only Goodeid lacking embryonic trophotaeniae. But, some years ago, while I was supervising an undergraduate research project on *Ameca splendens* at the University of Bath, I examined some newly born *Ataeniobius toweri* fry under a scanning electron microscope and became convinced that the fry's anal aperture was surrounded by trophotaenial tissue. The grape-like clusters are virtually identical to, although smaller than, the mucus-covered trophotaenial tissue that can be observed in mid-term *Xenoophorus captivus* embryos.

It is known that the size of Goodeid trophotaeniae is at its peak during mid-gestation, gradually shrinking as parturition approaches. Therefore, when a Goodeid fry is born, its trophotaeniae are considerably smaller than they were some two weeks earlier. This being the case, one would expect to find that, if *Ataeniobius toweri* has trophotaeniae at all, they would be larger in mid-term embryos than in new-born fry. The foundations for a research programme to investigate this possibility were actually

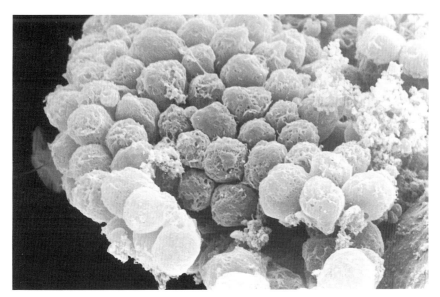

Plate 12   Scanning electron micrograph of a small section of a *Xenoophorus captivus* embryo's trophotaenial tissue. (*John Dawes*; reproduced by kind courtesy of the Centre for Electron Optics, University of Bath, Avon, England)

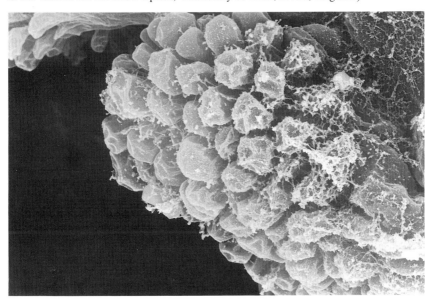

Plate 13   The similarity between this scanning electron micrograph of the tissues surrounding the anus of a newly born *Ataeniobius toweri* fry and the equivalent one from a *Xenoophorus captivus* embryo is very marked, indicating that *A. toweri* may, after all, possess trophotaeniae. (*John Dawes*; reproduced by kind courtesy of the Centre for Electron Optics, University of Bath, Avon, England)

laid down, in association with a researcher in another institution, before I
left my academic post to follow a full-time freelance career. Unfortunately,
the limited research funds originally available were re-directed shortly
afterwards, so the programme was never completed.

This is not the end of the story, though. Some time ago, Derek Lambert,
a British aquarist who specialises in livebearers and who has bred *Ataeniobius toweri* in some numbers, discovered that one of his females had
aborted halfway through gestation. The embryos were not fully developed,
of course, but they did have what appeared to be distinct trophotaeniae
extending from their vent. He therefore examined them under a light
microscope and arrived, quite independently, at the same conclusion as I
had done – that *Ataeniobius* does have distinct trophotaeniae, and that
these appear to be of the rosette-type found in *Goodea atripinnis*, a species
of Goodeid characterised by possessing very small trophotaeniae.

At the time of writing, Michael Smith is currently involved in a Goodeid
research programme at the American Museum of Natural History in New
York which could confirm the above conclusions beyond further doubt. If
that proves to be the case, then some of the obstacles and questions posed
by *Ataeniobius* regarding its reproductive biology and its relation to the
other Goodeids are likely to be resolved.

This may well result in a re-classification of *Ataeniobius* itself, or of the
Goodeids overall. But, in the absence of such a re-classification, I have
chosen to adopt Parenti's 1981 proposal, assimilating all the Goodeids
within the single subfamily Goodeinae but without further subdivisions
into tribes or their equivalents.

Hemirhamphidae

According to certain classifications, including that found in Nelson (1984),
the Halfbeaks belong to the family Hemirhamphidae, which, in turn,
belongs to the superfamily Exocoetoidea of the suborder Exocoetoidei of
the order Cyprinodontiformes.

While not denying the similarities between the Halfbeaks and their
nearest relatives, the Garfishes, Needlefishes and Flying Fishes, Rosen and
Parenti (1981) concluded that their relationship to the Cyprinodontiforms
(which include the Poeciliids and Killifishes) was not so close as to warrant
their inclusion in the same order. Accordingly, they set up a new order, the
Beloniformes, for these rather specialised fishes.

They then divided the Beloniform fishes into two suborders. The first,
the Adrianichthyoidei, contains the family Adrianichthyidae, which assi-
milates the Medakas – the Oryziidae – and the Horaichthyidae. The
second, the Exocoetoidei, is assigned two superfamilies – the Exocoetoidea (the Flying Fishes and Halfbeaks) and the Scomberescoidea (the
Needlefishes). Within the Exocoetoidea, the difference between the Flying
Fishes and Halfbeaks is recognised and each is assigned to a separate
family, the Flying Fishes to the Exocoetidae and the Halfbeaks to the
Hemirhamphidae. Finally, the Hemirhamphidae is subdivided into
subfamilies to take account of similarities and differences between the

genera. The better-known aquarium Halfbeaks such as *Dermogenys*, *Hemirhamphodon* and *Nomorhamphus* are placed within the subfamily Hemirhamphinae, while lesser-known fish such as *Zenarchopterus* are placed in the subfamily Zenarchopterinae.

TABLE VII   CLASSIFICATION OF HALFBEAKS (LARGELY ACCORDING TO ROSEN AND PARENTI, 1981)

Order	Beloniformes
Suborder	Exocoetoidei
Superfamily	Exocoetoidea
Family	Hemirhamphidae
Subfamily	Hemirhamphinae
Genera	*Dermogenys*, *Hemirhamphodon*, *Nomorhamphus*
Subfamily	Zenarchopterinae
Genus	*Zenarchopterus*

The classification adopted in this book is that of Rosen and Parenti (1981), which is accepted by, among others, Manfred Meyer *et al.* in their German-language book on livebearers, *Lebendgebärende Zierfische* (see page 29 for full reference).

Plate 14   Halfbeaks like *Dermogenys pusillus* are now regarded by many workers as belonging to the subfamily Hemirhamphinae of the family Hemirhamphidae. (*David Allison*)

Conclusion

The classification of livebearers, whether in the restricted sense of aquarium species, or in the much wider one embracing *all* fish that exhibit facultative or obligate viviparity, is a highly complex subject that is always in a state of flux and within which there is always room for debate and revision.

TABLE VIII   SUMMARY OF LIVEBEARERS FEATURED IN PART III

Family	Subfamily	Representative Species
1. Characidae (Nelson)*	Characinae (Nelson)	*Corynopoma riisei*
Characidae (Géry=Nelson's Characinae)	Glandulocaudinae (Géry)	
2. Oryziidae (Nelson)	—	*Oryzias latipes*
Adrianichthyidae (Rosen & Parenti)		
3. Rivulidae (Parenti)	—	*Cynolebias brucei* *C. melanotaenia*
4. Poeciliidae (Parenti)	Poeciliinae (Parenti)	*Alfaro cultratus* *Gambusia* spp *Poecilia* spp *Poecilia (Lebistes) reticulata* *Xenodexia ctenolepis* *Xiphophorus* spp
5. Goodeidae (Parenti)	Goodeinae (Parenti)	*Ameca splendens* *Characodon lateralis* *Goodea atripinnis* *Skiffia francesae*
6. Anablepidae (Parenti)	Anablepinae (Parenti)	*Anableps anableps* *Jenynsia lineata*
7. Hemirhamphidae (Rosen & Parenti)	Hemirhamphinae (Rosen & Parenti)	*Dermogenys pusillus* *Hemirhamphodon* spp *Nomorhamphus* spp

* The names in brackets represent the authors whose work has been quoted in connection with the relevant classifications in this chapter. They are not necessarily the originators/creators of the families/subfamilies in question, but their works can serve as useful starting points for further literature searches and as sources of reference to the relevant original publications.

Several fish that do not normally feature in definitions of livebearers have been included in this book for the sake of completeness, and on the basis that they exhibit facultative viviparity to a greater or lesser extent. The species in question – two Killifishes, the Medaka and the Swordtail Characin – have been brought within the boundaries of the definition on the basis that they differ only, or largely, in degree from the facultative livebearing exhibited by *Tomeurus gracilis* – a member of the Poeciliinae, the subfamily containing the classic livebearers.

The stretching of the boundaries of the definition to embrace these fish means that the classification of livebearers that I have adopted will not find universal acceptance. However, by adopting this approach – even accepting that the debatable species are on the fringe of what could conceivably be considered as viviparity according to Wourms' 1981 criteria – I hope to present a more complete picture of those aquarium species that we can regard, with or without qualification, as exhibiting *some* degree of livebearing. The families and their representatives covered in the species section later are summarised in Table VIII opposite.

Further Reading

Géry, Jacques, *Characoids of the World,*   Tropical Fish Hobbyist Publications, Inc., p. 672 (1977).

Jacobs, Kurt, *Livebearing Aquarium Fishes,*   Tropical Fish Hobbyist Publications, Inc., p. 495 (1971).

Miller, Robert Rush and Fitzsimons, John Michael, '*Ameca splendens*, a New Genus and Species of Goodeid Fish from Western Mexico with Remarks on the Classification of the Goodeidae',   *Copeia*, 1, pp. 1–13 (March 8, 1971).

Nelson, Joseph S., *Fishes of the World* (2nd Edition),   Wiley-Interscience, p. 523 (1984).

Parenti, Lynne R., 'A Phylogenetic and Biogeographic Analysis of Cyprinodontiform Fishes (Teleostei, Atherinomorpha)',   *Bulletin of the American Museum of Natural History*, Vol. 168, Article 4 (1981).

Parenti, Lynne R. and Rauchenberger, Mary, 'Systematic Overview of the Poeciliines' in *Ecology and Evolution of Livebearing Fishes,*   Gary K. Meffe, Franklin F. Snelson, Jnr., *et al.*, Simon & Schuster, p. 450 (1989).

Rosen, Donn E. and Bailey, Reeve M., 'The Poeciliid Fishes (Cyprinodontiformes), their Structure, Zoogeography and Systematics',   *Bulletin of the American Museum of Natural History*, Vol. 126, Article 1 (1963).

Rosen, Donn E. and Parenti, Lynne R., 'Relationships of *Oryzias*, and the Groups of Atherinomorph Fishes',   *American Museum Novitates*, 2719, pp. 1–25 (1981).

Form and Colour

The Guppy has been firmly established among the top ten most popular aquarium fishes for longer than most people care to remember. Yet, it was only in the mid-1960s that the explosion in colour and finnage variations that we know today really began. Since then, ever-more-ambitious and extreme body and fin configurations have appeared, to the extent that the wild-type Guppy is now a little-known and little-seen fish in the general aquarium hobby.

The trouble with such situations, however colourful or spectacular the resultant fish may be, is that people soon start to lose touch with the *real* fish – the one that's made all these large, colourful, over-finned mutants possible.

It is estimated that Singapore produces about forty different varieties of Guppy. Florida-based farmers are also responsible for an impressive selection, while Japan, Israel, some European countries (including Germany and the United Kingdom), Russia and even a few African countries (whose fish are often particularly sturdily built) all produce larger or smaller numbers of Fancy Guppy.

The one thing that all these producers have in common is that their Guppies are long-finned, large-bodied fish – beautiful in their own way, but far-removed from their tiny, multi-coloured, short-finned ancestors.

I have selected Guppies as an example of diversity of form in livebearers because, when we talk about form, we need to establish exactly what it is we are referring to, and there are few better examples around of a fish whose true form is seen by so many different people in so many different ways.

To many, form in Guppy terms is represented by a wide-tailed, often slightly hunch-backed fish with a flowing, pennant-like dorsal fin. This profile has tended to become *the* Guppy shape over the years. However, this image of a Guppy is about as far removed from the really true form as one could imagine.

This wild-type Guppy is a colourful, short-finned fish in which the males are sometimes no more than around 2 cm (0.75 in) long, the females being considerably larger. In the males, the body shape is almost that of a slim cylinder, pointed at the front end, and rounded at the back. Yet, even within this simple wild-type pattern, there are variations, with caudal-fin extensions resulting in morphs such as the Pintail, Bottom Sword, Double Sword and others.

Plate 15   A spectacular Guppy (*Poecilia reticulata*) male – one of numerous man-made varieties of this remarkable fish. (*Bill Tomey*)

These minor variations clearly indicate a certain degree of plasticity within the genetic make-up of Guppies and it is this that has been worked on and developed to produce today's bewildering array of varieties.

Despite this inherent propensity for variation, all wild-type Guppies are short-finned – a clear indication that only the best forms are capable of surviving in the rough, uncompromising world of nature. For wild Guppies, like all other truly wild fish, have evolved in tune with their environment, undergoing severe, but totally natural, selective pressures every second of every minute of every hour of every individual's short and hectic life. The result is a species of fish whose members are difficult to spot from above, whose males have dazzling colours which they use both to impress females and discourage rivals, and whose turn of speed over short distances, coupled with their small size, allows them to make lightning-fast, effective getaways, especially in thickly vegetated areas.

Females are drab-coloured; they do not need the bright colours of the males. In fact, their drabness offers them a considerable degree of protection – an especially important factor when they are carrying developing young inside. They, too, like the males, are difficult to spot from above and are capable of blurringly fast spurts when alarmed.

These characteristics are, of course, not restricted to Guppies. Numerous livebearers, mainly Poeciliines, have similar attributes, though few can boast the Guppies' exceptional colours. The fact is that a cylinder- or torpedo-shaped body is excellent, both for chasing food and escaping predators, and, since many livebearers share a similar lifestyle, evolution has moulded them accordingly.

Diversity of 'Surface' Features

Other features that many livebearers share include an upturned, straight-edged mouth, a more-or-less flat top to the head (continuous with an equally straight-edged back) and a well-set-back dorsal fin. Further, a notional line drawn from the centre of the eye to the centre of the caudal peduncle – the base of the caudal, or tail, fin – is usually almost parallel, or at a very shallow angle, to the top edge of the back.

The upturned mouth is indicative of a life spent predominantly, but not necessarily exclusively, near the surface. The diet of many Poeciliine species consists largely of flying insects that fall on to the surface of the water, or aquatic ones that are found on, or rise to, the surface for air, e.g. mosquito egg rafts (cases), larvae, pupae and hatching and egglaying adults.

An upturned mouth, combined with a slim body shape with the dorsal fin set well back and a caudal fin whose longitudinal axis is parallel (or almost so) to the air/water interface, allows the fish to swim at high speed just under the surface without having to invest in a large expenditure of energy. This obviously makes such fish ideal hunters of the upper reaches of the water, to which many species adhere as if held there by an invisible magnet. This tendency to remain within the top few inches of the water column has led to many species of livebearers being referred to as 'surface swimmers' or 'top minnows'. Some even have common names linking them with the insects that are found in these habitats, e.g. the Mosquito Fish (*Gambusia affinis*).

Mouth Diversity

Most surface species, despite their overall preference, will also swim, either occasionally or habitually, in the lower reaches. Here, their straight-edged mouths are also put to good use to scrape encrusting algae, with their resident microfauna, from plants, submerged branches, rocks, river banks etc. A moveable lower jaw, armed with numerous small teeth, closing against a more-or-less rigid straight edge, is a perfect combination for this job, and those species that exhibit this type of behaviour (notably the Mollies) have this kind of mouth structure.

Not all surface swimmers have all the above characteristics, though. For instance, the Piketop Livebearer (*Belonesox belizanus*) is an out-and-out predator with no interest in vegetation whatsoever – other than in any prey that might be hiding in it!

Consequently, while it shares some of the surface-type characteristics of many of its nearest relatives, its mouth and teeth are totally different.

Plate 16   The Piketop Livebearer's dentition bears the unmistakable trademark of the supreme predator. (*Bill Tomey*)

Gone is the small-toothed, straight-edged mouth with a highly flexible lower jaw. Instead, the mouth of the Piketop is almost beak-shaped and is armed with a fearsome set of very sharp pointed teeth. Coupled to this is a finely developed predatory instinct. The lower jaw, while highly mobile along a vertical axis, does not possess the soft, flexible, fleshy lips of the grazers and mosquito-eaters. Neither does it possess the cutting, dog-like teeth of that other famous predator, the Piranha. Instead, its numerous teeth are needle-shaped – ideal for grabbing and holding on to prey, which is then swallowed whole. With females growing up to around 18 cm (7 in), and males to only around 10 cm (4 in), many a small amorous male has ended up as a tasty meal for a hungry non-receptive female. This, in human terms, is a fish without scruples – but a fantastic hunting machine nevertheless.

The Halfbeaks (subfamily Hemiramphinae) also have predatory habits, though they are not as acutely developed as in the Piketops. Of all the fish dealt with in this book, the Halfbeaks are perhaps the most surface-bound of the lot (with the exception of *Anableps*). Everything about their body shape is geared to this type of existence – their eyes are located near the top of the head, the mouth is inflexibly directed upwards, there is no straight-edge to it (this fish is not a grazer by any stretch of the imagination), the dorsal fin is set even further back than in Poeciliine surface swimmers, and the eye/caudal-peduncle notional line is even more nearly parallel to the water surface than in the other species. Halfbeaks are surface predators par excellence.

Unique Eye Features

The most remarkable features of the Four-eyed Fish (*Anableps*) are, of course, their eyes, of which they only have two, despite their very descriptive, but slightly misleading, name. The fantastic thing about these eyes is that they can see above and below the water surface at one and the same time. The secret lies in the eye lens and in a unique horizontal strip of pigmented tissue, which effectively cuts each eye into two along the water line.

But before it can see both above and below water level, the Four-eyed Fish has to solve the problem created by the different refractive indices (bending powers) of air and water. Water, which is denser than air, has the higher refractive index of the two. This explains why, say, a straight stick held half immersed in water will appear to bend away from the person holding it. The kink occurs at the air/water interface.

The best eye lens for seeing in air is one shaped like a long, slim oval – that is, with slightly convex sides that are aligned more or less at right angles to the incoming light. Underwater, though, this arrangement is next to useless. A stronger lens is needed – something with more or less the sort of curvature possessed by a sphere. Fish eye lenses are, in fact, this shape, so fish can see well underwater, but very poorly in air.

This, then is the challenge facing Four-eyed Fish, in which half of each eye is above water and half below. The solution that evolution has come up with is quite fantastic. The shape of the eye lens in these fish has changed in such a way that the surface facing upwards, into the air, is slightly convex, like a human lens. But the ends of the lens, which look downwards into the water, are rounded – the ideal shape for underwater focusing. The Four-eyed Fish has evolved, in fact, a lens that can focus light rays coming in both from the air and the water simultaneously – quite a remarkable feat.

However, the modifications do not end there. Four-eyed Fish have *two* retinas in each eye, instead of the more normal one. The dorsal retina, the one in the top half of the eye, receives light rays coming in from the water below, once these have been focused by the two round ends of the lens. The ventral retina receives the rays coming in from above, once they have been focused by the weaker, flatter and longer sides of the lens.

The final refinement comes from the pigmented band mentioned earlier. This band is made up of two separate layers. The bottom, or internal, one – that is, the one lying in contact with the surface of the eye – consists of two flaps of tissue that grow inwards from either side of the iris. These two flaps meet in the centre of the pupil, but, even though they touch, their tissues don't fuse. Lying on top of these flaps is a thin layer of pigmented cells that prevent light rays from straying across, thus improving the separation between the air and water halves of the eye.

As a piece of biological engineering, the eyes of a Four-eyed Fish are nothing short of a masterpiece, shaped by the powerful forces of natural selection and capable of effortlessly performing what would be an otherwise impossible job.

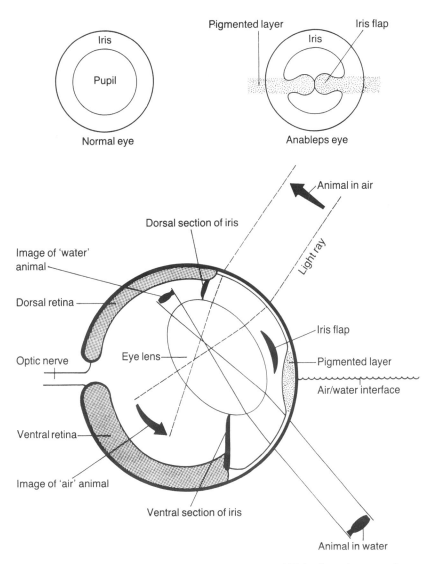

Fig. 8 The remarkable eye structure of the Four-eyed Fish allows it to see above and below the water surface simultaneously.

Gonopodial Diversity

Turning to the fins, all show some diversity of form, but none more so than the gonopodium.

From a distance, one rod-like gonopodium looks very much like the next. The only easily discernible differences are in overall length. But examine a gonopodium under even the weakest of light microscopes and a totally different situation becomes immediately apparent.

For a start, the featureless rod, is not a rod at all. It is a series of fin rays aligned behind one another, just as in a normal fin. But there the similarities end. Some of the rays, particularly the third, fourth and fifth, carry all sorts of adornments, from simple comb-like spines to hooks, claws and blades.

These modifications, while exhibiting a certain overall uniformity in the rays that are affected, nevertheless show very distinct differences between genera and between species within a genus. In fact, the gonopodium is recognised as being the single most significant discriminatory feature between closely related Poeciliine livebearers.

It is therefore possible to distinguish between the gonopodia, say, of a *Xiphophorus* and a *Poeciliopsis* or a *Poecilia*. And within the genus *Poecilia*, there is a subgeneric Limia-type gonopodium, distinct from a subgeneric Poecilia-type gonopodium. Then, within each genus, individual species have their own unique pattern of gonopodial adornments.

It is not only the internal arrangements of the bones of the gonopodium that exhibit diversity, though. The overall length also varies from species to species.

An interesting aspect of this diversity of length is its overall relation to breeding behaviour.

Typically, species of *Poecilia* (Mollies and their relatives) and *Xiphophorus* (Swordtails, Platies and their closest relations) and many other Poeciliines have very short gonopodia, while others, among them *Heterandria*, *Phallichthys*, *Poeciliopsis* and *Belonesox* have rather long gonopodia. Clearly, the shorter the gonopodium, the closer a male needs to get to a female in order to make contact and effect sperm transfer. Should a female be uncooperative, then successful mating would be next to impossible.

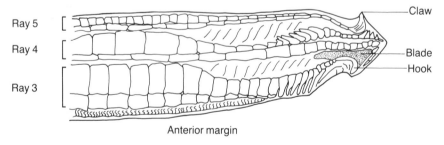

Fig. 9 The gonopodium of the male Poeciliid (this is a Swordtail, *Xiphophorus helleri*) carries a number of structures that are unique for each species.

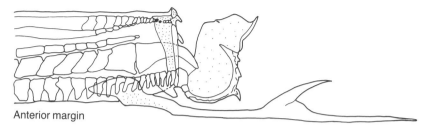

Anterior margin

Fig. 10 The extraordinarily complex gonopodial tip of *Tomeurus gracilis*. (After Rosen, 1952)

What we tend to find, therefore, is that species with short gonopodia exhibit a relatively well developed degree of cooperation between the sexes.

It may sometimes appear, at first glance, that a male rushes in, mates with an unsuspecting female and then makes his escape before she is even aware that copulation has taken place. This, at least, is the picture sometimes painted within the aquarium hobby. However, this version of Poeciliid mating behaviour is way off the mark. The fact is that, in numerous short-gonopodium species, females respond quite positively to the males' advances. These aproaches are often accompanied by a considerable amount of fin expansion, intense body wagging (courtship dances) and other associated activities in full sight of the females. Acceptance of a male's advances is often signalled by head-shaking on the part of the female, followed by a dipping of the anterior part of her body (with a

Plate 17 Many Poeciliids, such as Platies (*Xiphophorus maculatus*), have short gonopodia. In this Red Wagtail male, the gonopodium – a thin, black, abdominal appendage held close to the body – can be clearly seen. (*Ross Socolof*)

Plate 18   The extremely long gonopodium of this *Phallichthys fairweatheri* male can be fully appreciated in this photograph. (*Dennis Barrett*)

corresponding raising of the posterior half), which thus renders her genital aperture more accessible to the male once he adopts the correct mating position. As the male swings the gonopodium forward to make contact, the female will often fold her anal fin out of the way, helping matters even further.

If there is such a thing as non-cooperation, then it is to be found among those species in which the males have long gonopodia and do not need to get so close to a female. What we often find in these species is a form of stalking of a female by a male intent on mating with her.

Other Types of Male Reproductive Organs

The copulatory organs of livebearers show a considerable diversity in design. In the Poeciliines, despite some variations and some specific adornments, the underlying design is that the fin rays are aligned one behind the other as in any other normal fin. But this is only one of many general designs.

In Anablepines, for instance, the gonopodium is formed into a tube through which the sperm travel during mating. This is quite different to the temporary channel or canal formed by the gonopodial rays of most male Poeciliines as the fin is swung forward. At the base of the Anablepine gonopodium, the rays are arranged in a line as in Poeciliines. However, as we move down the fin towards the tip, the rays twist around each other to form a sort of circle that is then overlain by the fleshy tissues of the tube proper, itself a continuation of the sperm duct.

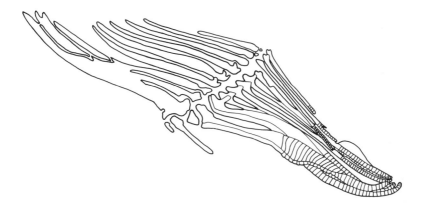

Fig. 11 In *Anableps*, the gonopodial rays are twisted round each other – a characteristic of this subfamily. (After Turner, 1950)

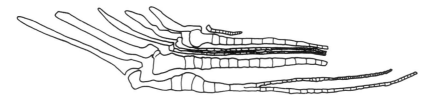

Fig. 12 Halfbeak males have highly modified anal fins known as andropodia.

In Halfbeaks, the anal fin of the males looks superficially like a Poeciliine or Anablepine gonopodium. However, closer examination shows very significant differences. These are so typical of Halfbeaks that the anal fin is often referred to as an andropodium (a term coined by Brembach in 1976) rather than as a gonopodium. Characteristic features of an andropodium include a unique knee-like structure, called a geniculus, situated approximately halfway down the second ray of the fin. In addition, this ray possesses two laterally spread spines (the tridens flexilis) with a spiculum (a part of the ray itself) situated between them. Further, the skin between the third and fourth rays forms a pouch-like structure, known as the physa.

A totally different situation is found in the Goodeinae. These remarkable livebearers possess no such sophisticated structure as a gonopodium or andropodium. Their male anal fin modifications are restricted to a bunching together and shortening of the first five to seven rays. This produces a characteristic notch that plays a crucial role in sperm transfer by forming a pocket when the front part of the male's anal fin, or spermatopodium, is pressed against the female's genital opening. At that point, an internal ring-like muscular organ contracts and produces a spurt of sperm packets (spermatozeugmata) which are more or less forcefully injected into the female.

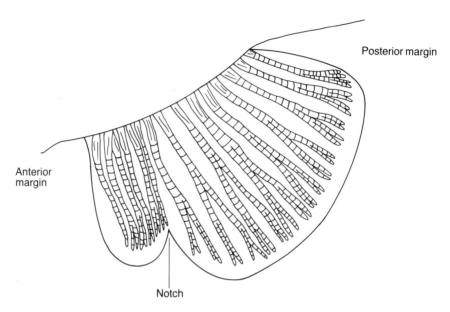

Fig. 13 The notched anal fin of the male Goodeid is known as a spermatopodium.

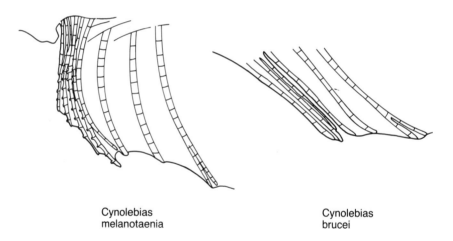

Cynolebias
melanotaenia

Cynolebias
brucei

Fig. 14 Both *Cynolebias brucei* and *C. melanotaenia* have notched anal fins that are believed to play a role in sperm transfer. (Based on L. Parenti, *Bull. Am. Mus. Nat. Hist.*, Vol. 168, 1981)

The absence of an elongated copulatory organ in these fish dictates that males and females need to come into very intimate contact with each other for successful mating to take place. This, in turn, makes it essential for a female to cooperate fully with a courting male, something that can only be achieved if both parties can convey and receive the appropriate sexual messages. This is accomplished by courtship displays that are generally longer, more complex and more sophisticated than those exhibited by other livebearers. Understandably, therefore, Goodeids have developed an enthusiastic following among those aquarists who have been fortunate enough to observe their dazzling courtship dances.

Finally, on the subject of anal fin diversity, I must mention two Killifish that exhibit facultative viviparity – *Cynolebias brucei* and *C. melanotaenia*. These species have condensed anterior rays in their fins, like those of Goodeids but not quite as bunched or notched. They do apparently serve the same purpose, however, but no exact details regarding the way in which they achieve sperm transfer appear to be available.

In *Oryzias* and *Corynopoma*, there are no easily discernible male anal fin modifications (*Corynopoma* males have tiny hooks on some of the rays). Nevertheless, in the latter species, the long paddle-like structure on the male's gill covers may be a reproductive feature, since it appears that it is essential for a receptive female to bite its fleshy tip for successful mating to take place. Again, few documented details are available regarding either the precise function of this structure or the actual process of mating.

Male Paired Fins in Poeciliines

One of the several advantages of gonopodial diversity among Poeciliines is that it helps prevent species from hybridising in the wild, even should their ranges overlap. Behavioural differences also help, but even if these were to break down, the specific, physical differences in gonopodial tip modifications make interbreeding relatively difficult.

In the close confines of an aquarium, however, the artificial conditions can lead to deviations from the normal, with a possible breakdown of isolating barriers. Even here, though, hybridisation is generally restricted to species that share certain common characteristics (in addition to close genetic similarities, of course). We are therefore most unlikely to encounter effective cross mating between, say, a *Heterandria formosa* male and a Guppy female. *Heterandria* is an external depositor – the spermatozeugmata are trapped by a large foliate genital pad surrounding the female's genital orifice. Guppies, on the other hand, are internal depositors, with the spermatozeugmata being inserted into the female's genital aperture rather than being laid on its external surface.

There are other differences, too. *Heterandria formosa*, for example, is a species that possesses a long gonopodium, while Guppy males have short gonopodia, a feature they share with their close relatives, the Mollies, Limias and others. The possession of a long gonopodium is, as we have seen, associated with a mating strategy that is virtually, or totally, devoid of cooperation between the sexes. The males of these species orientate their gonopodia visually, darting in and making no more than momentary contact, often with an unsuspecting female. These species are probably

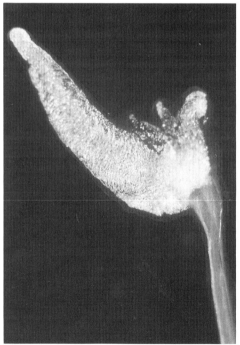

Plate 19 ABOVE The gill-cover 'paddle' that the male Swordtail Characin (*Corynopoma riisei*) in this photograph is holding away from his body may play a vital role in ensuring that the female (lower fish) attains the correct orientation for mating. (*Bill Tomey*)

Plate 20 A close-up of a *Corynopoma* 'paddle'. (*Bill Tomey*)

able to take aim visually, because the length of their gonopodium – usually reaching up to, or beyond, the eye when directed anteriorly – brings the tip within their field of vision, thus making it possible for them to aim their thrust without too much difficulty.

When it comes to species that possess a short gonopodium, the situation is quite different. Even at the point of maximum anteriorly directed extension, the gonopodial tip lies well behind the male's field of vision. There is no way, therefore, that such males can align themselves using purely visual cues. Yet, they are just as successful at mating as their longer-finned counterparts.

It is here that the fleshy hoods and sensory spines that such species have around their gonopodial tip comes into play. Equally importantly, though, modifications in the pectoral or pelvic fins of these males also play a crucial role by helping the gonopodium during its forward mating thrust. In *Gambusia* species, for instance, some of the anterior rays of the pectoral fins of males (starting at the second or third ray and ending at the fifth) are thicker than the rest and form a sort of downwardly directed notch when these fins are extended. As the gonopodium is swung forward and sideways in a mating attempt, the pectoral fin on the same side of the body as the swing is extended. Ray 4 of the gonopodium has an elbow-like joint that projects upwards as the swing nears its full forward extension. At that point, contact is made between the elbow and the notch. Whether they lock into each other or not is not clear, but the overall effect is that their cooperative action helps stabilise the gonopodium and enhances the chances of achieving successful sperm transfer.

In Mollies, Guppies, Swordtails, Platies and some of their close relatives this stabilising role is taken over by the pelvic, rather than the pectoral, fins. In *Xiphophorus* (Swordtails and Platies), males have a 'fleshy appendage ... along [the] distal third of [the] first short, unbranched ray of the pelvic. The second and third rays are somewhat prolonged' (Rosen and Bailey, 1963). In the genus *Poecilia* (Mollies, Limias and Guppies), the pelvic fins of adult males have a 'variably developed fleshy swelling distally on ray 1 and a long, fleshy, finger-like extension of [the] tip of ray 2. [Further] ray 2 [is] thickened subdistally, in many cases joining [a] distinct bony prominence that may enter or make contact with [the] terminal swelling on ray 1; rays 1 and 2 [are] joined by dense connective tissue and [are] usually separated from rays 3 to 5 by [a] distinct notch' (Rosen and Bailey, 1963).

Perhaps the most extreme paired-fin modifications found in Poeciliines are those possessed by the intriguing *Xenodexia ctenolepsis*. So different are these modifications that this member of the subfamily Poeciliinae has a tribe, the Xenodexiini, all to itself (Parenti and Rauchenberger, 1989).

Xenodexia differs from all its relatives in that males are distinctly asymmetrical – in other words, a notional vertical line cutting the fish into two, results in unequal halves. Some other Poeciliines, e.g. *Poeciliopsis* and *Carlhubbsia*, which have long gonopodia, are also somewhat asymmetrical in the sense that some of the gonopodial rays are more developed on one side than on the other, resulting in a greater or lesser degree of right-

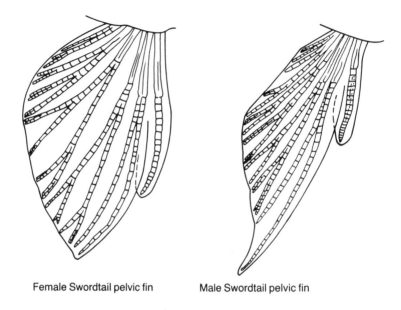

Female Swordtail pelvic fin   Male Swordtail pelvic fin

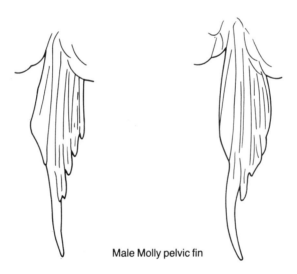

Male Molly pelvic fin

Fig. 15   The pelvic fins of male Mollies and Swordtails carry modifications in the form of notches or extended rays which aid in the correct alignment of the gonopodium during mating. (Molly fins drawn after Rosen and Bailey, 1963)

handedness or lefthandedness, but in *Xenodexia* the asymmetry is carried two significant stages further. The right pectoral fin of the males carries a fleshy modification, usually referred to as a clasper, while the right pelvic fin carries 'a thickened fleshy ridge along the ventromesial [middle/ventral] edge of the proximal [closest to the body] third of the outer ray' (Rosen & Bailey, 1963, based on Hubbs, 1950).

Xenodexia also differs from other Poeciliines in that its extremely long gonopodium is tubular and therefore, in this sense at least, more akin to that found in the Four-eyed Fish (*Anableps* spp) and the One-sided Live-bearer (*Jenynsia lineata*), belonging to the subfamily Anablepinae.

The right pelvic and pectoral fin modifications of *Xenodexia* males are assumed to perform a stabilising function when the gonopodium (which is approximately 45% the length of the whole body) is swung forward in a mating attempt. Curiously, though, *Xenodexia* males do not possess the elaborate hooks and claws found on the gonopodial tip of other species such as *Xiphophorus helleri*, that also inhabit fast-flowing waters.

It is believed that the highly developed mechanism of the pectoral fin somehow clasps the anal fin of the female during copulation. How this is achieved, how the clasper also helps to stabilise the gonopodium, or how the thickened fleshy ridge of the right pelvic fin fits into the overall picture has not really been worked out. However, an expedition mounted by Ross Socolof and Jaap-Jan de Greef in 1989 resulted in substantial collections of live specimens of *Xenodexia* that may well yield some answers and give us some interesting insights into the courtship behaviour of this amazing livebearer.

Plate 21   The right pectoral fin of the *Xenodexia ctenolepis* male (upper fish) carries a clasper that is believed to help stabilise the gonopodium during copulation. (*Ross Socolof*)

Conclusion

Livebearers exhibit a considerable degree of diversity, both in the overall shape of their bodies as well as in their constituent parts – mouths, eyes, finnage, etc. Each permutation is finely tuned to the survival needs of the species in question.

Wild populations are considerably more conservative in body form and coloration than cultivated ones, the latter only being capable of long-term survival in the artificial confines of aquaria. Nevertheless, the existence of fancy varieties is only made possible because of the reservoir of genetic variability that the species concerned possess.

Without variation there can be no evolution, and although the external manifestation of this genetic diversity or polymorphism is held in check by the pressure of natural selection, it nevertheless exists, ready to spring forth should a change in environmental conditions favour the expression of one characteristic or other. Many livebearers are therefore perfectly placed to exploit a wide range of habitats and diets and are, consequently, survivors par excellence, their external and internal diversity equipping them admirably to withstand the many and changing challenges that nature constantly confronts them with.

Further Reading

Chambers, James, 'The cyprinodontiform gonopodium with an atlas of the gonopodium of the genus *Limia*', *Journal of Fish Biology*, 30, pp. 389–418 (1987).

Constantz, George D., 'Reproductive Biology of Poeciliid Fishes' in *Ecology and Evolution of Livebearing Fishes*, Gary K. Meffe, Franklin F. Snelson, Jnr., *et al.* Simon & Schuster, p. 450 (1989).

Iwasaki, Noboru, *Guppies (Fancy Strains and How to Produce Them)*, Tropical Fish Hobbyist Publications, Inc., p. 144 (1989).

Jacobs, Kurt, *Livebearing Aquarium Fishes*, Tropical Fish Hobbyist Publications, Inc., p. 495 (1971).

Parenti, Lynne R. and Rauchenberger, Mary, 'Systematic Overview of the Poeciliines' in *Ecology and Evolution of Livebearing Fishes*, Gary K. Meffe, Franklin F. Snelson, Jnr., *et al.*, Simon & Schuster, p. 450 (1989).

Rosen, Donn E. and Bailey, Reeve M., 'The Poeciliid Fishes (Cyprinodontiformes), their Structure, Zoogeography and Systematics', *Bulletin of the American Museum of Natural History*, Vol. 126, Article 1 (1963).

Schröder, Johannes Horst, *Genetics for Aquarists*, Tropical Fish Hobbyist Publications, Inc., p. 128 (1976).

Scott, Peter W., *A Fishkeeper's Guide to Livebearing Fishes*, Salamander Books Limited, p. 117 (1987).

Socolof, Ross, 'The Quest for *Xenodexia*', *Aquarist & Pondkeeper* (Nov. 1989).

Turner, Bruce J. (ed), *Evolutionary Genetics of Fishes*, Plenum Press, p. 636 (1984).

Sex Determination

General Genetic Principles

Every living organism is born with an in-built set of characteristics that identify it as a member of a particular species. These characteristics may become modified as the organism grows, but however extreme they may eventually appear, they will nevertheless remain within certain imposed limits.

A new-born Guppy, for example, will change dramatically as it grows. Yet, no matter how different it may appear as an adult when compared to what it looked like at birth, it will still be identifiable as a Guppy, as a member of the species *Poecilia reticulata*. In other words, it will develop according to a specific set of rules, dictated by some form of internal control system that will not allow it to deviate too far from the Guppy blueprint.

These internal commands originate in the genes, which are gathered together into strand-like structures called chromosomes. The chromosomes are, in turn, found inside the nucleus of each living cell. The moment an egg is fertilised, it ends up with a full set of encoded genetic instructions that will allow it to develop into a complete, identifiable, functional member of the species concerned. Thus, a Guppy egg will always develop into a Guppy, a Swordtail egg into a Swordtail, and so on.

At the gene level, nature does not divide the total genetic complement up depending on which cells are destined to form nerves, eyes, muscle or anything else. *Every* cell has a complete blueprint. Basically, the final destiny of a particular cell is determined by genetic switches that are either 'on' or 'off'. Therefore although, say, cells 'A' and 'B' are both capable of developing into nerve cells, they will not do so unless their 'nerve switch' is 'on'. In the end, each cell in a developing embryo will have a number of genetic switches 'on' and a number 'off'. Failing this, all cells would develop in exactly the same way – hardly a viable method of putting together as sophisticated and complex an organism as a fish.

Taking the matter a stage further, it can be seen that a certain minimum number of genetic criteria need to be satisfied before an egg can end up as a fully formed individual of a particular species. For a start, an egg needs the genetic equipment necessary to develop into a *fish* egg, as opposed to an amphibian one, or a reptile one, or whatever. Coming down the scale, examples of other basic criteria that need to be satisfied are those differentiating between a cartilaginous fish and a bony one, or a goby and a barb, or a Goodeid and a Poeciliid, or an *Ameca splendens* and a *Characodon lateralis*, and so on.

The further down the line one travels, the more precise the genetic instructions need to be, so that, even within a genus, or within a single species, an egg needs the appropriate commands. Taking a concrete example, a fertilised *Gambusia affinis* egg needs genetic commands to allow it to develop into a bony fish that belongs to the Cyprinodontiformes. Within that general message, there have to be more precise instructions that tell the egg to develop into a Poeciliid. Then there are those commands that will direct development along the Poeciliine path (as opposed to, e.g. the Aplocheilichthyine one). Finer tuning will result in the egg developing into a fish that belongs to the genus *Gambusia*. Even more precise genetic commands will switch on the genes that will allow the fish to become a fully functional, identifiable member of the species *Gambusia affinis*.

The commands don't end there, of course. In many ways, this is just the beginning, because the genes play a part in everything that an organism does, from the types of food it can digest, to the shape of its mouth, the maximum size it can attain, and even its sex.

Sex Determination in Fish

While the sex of a mammal can usually be relatively easily determined by looking at the chromosomes of a living cell, things are not quite as clear-cut when it comes to fish.

Certain species, such as *Geophagus brasiliensis*, one of the Earth Eaters (a Cichlid) and at least one Goodeid, *Allodontichthys hubbsi*, have clear-cut, identifiable sex chromosomes. However, in the majority of cases, the sex chromosomes are morphologically indistinguishable, although cytologically distinct, from the other, non-sex chromosomes – the autosomes. They are also morphologically indistinguishable between the sexes. This means that in most fish the male and female cells cannot be distinguished, as they can in mammals, by their possession of X or Y chromosomes.

This is because sex determination in fish is *polygenic* or *polyfactorial*. As both these terms indicate, numerous genetic interactions are at work when it comes to determining the sex of an individual fish.

First put forward by Winge in 1934, the polygenic theory allocates sex determination not only to those genes that may be located on the genuine sex chromosomes, but also to a host of others that occur on the autosomes. This applies both to species in which the sex chromosomes are identifiable and to those in which they are not. In the end, it is not just a question of whether a fish has identifiable sex chromosomes or not; it is more a question of the ratio of male to female factors determining the sex of an individual.

The sex chromosomes are assumed to carry superior sex genes that have greater male or female genetic power than the other male and female genes located on the autosomes. These male and female factors are often referred to as M and F determinants respectively. It is also believed that the strength of the M and F factors on the autosomes are about evenly matched. If this is the case, it would then be likely that they would cancel each other's influence out in the normal course of events. This, of course, is of no use whatsoever when it comes to tipping the scales one way or the other in terms of maleness or femaleness. What appears to happen is that the M and F factors within the sex chromosomes themselves determine which genetic switches are going to be turned on or off in the autosomes. If a predominance of male-type switches are turned on, then the fish in question will develop into a male, and vice versa. This may explain to an extent why environmental factors appear to play such a publicised role in sex determination in certain fishes.

The best-known examples are the well-documented sex reversals found in Wrasses and Clownfish. But statements regarding sex reversal in Poeciliids abound, particularly in the Swordtail (*Xiphophorus helleri*), although it has also been reported among Guppies (*Poecilia reticulata*) and others. Yet, despite all these reports, scientific confirmation of the occurrence of sex reversal is still lacking. Most of the scientific reports recorded so far have been found to be 'so poorly documented as to raise serious doubts whether any "sex reversals" really occurred' (Klaus D. Kellman, 1984).

Further investigations into other reports revealed that so-called reversed females which had, reportedly, given birth to fry prior to changing sex, had been kept with other females that were *known* to have produced fry and were known *not* to have changed sex. The conclusion reached was that the so-called females were, in fact, late-developing males.

It has also been shown that the anal fins of mature females of certain Poeciliine species (*Gambusia affinis* and *Xiphophorus maculatus*) are incapable of changing into fully functional gonopodia.

Elongated anal fin rays and extension of the lower caudal fin rays in Swordtail females belonging to the species *Xiphophorus helleri*, *X. signum* and *X. alvarezi* are two secondary male characteristics that have been documented. Although it has been found to occur predominantly in old females, this *arrhenoidy* is deemed to be genetically linked but subject to external influences. Internally, those masculinised females that have been examined have been found to have fully formed ovaries, with no sign of testicular tissue.

At the present time, therefore, true sex reversal in livebearers is not believed to exist.

There are also numerous, largely anecdotal and undocumented, accounts that sex ratios in certain species of livebearers, such as some of the Limias, are affected by increases in temperature during development – the higher the temperature, the higher the incidence of males. One fully documented example does exist, however, where *Poeciliopsis lucida* broods were observed to develop into a predominantly higher number of males as the rearing temperature was increased (Schultz 1986 see page 76).

Another environmental parameter said to affect sex determination is pH, with the percentage of females increasing in line with the values. This has been claimed both for the Swordtail (*Xiphophorus helleri*) and for Cichlids like the Kribensis (*Pelvicachromis pulcher*). In the former case, some experimental evidence exists (see page 77). In the latter case, I know of no fully documented scientific account.

Be that as it may, there can be little doubt that some environmental factors can have some influence in sex determination in fish, probably by controlling which of the M and F genetic switches are activated.

So far, I have referred to X and Y chromosomes and M and F determinants. It should be stressed here that the M and F factors are not complete chromosomes, just parts of them. We cannot therefore say that fish have X, Y, M and F chromosomes. What we can say is that fish possess X and Y sex chromosomes and M and F determinants, some of which are found on the sex chromosomes while others are found on the autosomes.

This is not the complete picture, though. A search through the literature will show references not only to X and Y sex chromosomes, but also to W and Z chromosomes in connection with sexual determination in some species (usually livebearers). A great deal of research and debate has surrounded these last two types of sex chromosomes. The general consensus of opinion appears to be that the Z chromosome, where it occurs (which is not in every species of fish), is, in fact, equivalent to the Y chromosome. However, no such evidence is forthcoming regarding the equivalence of W and X. Therefore, the conclusion that can be derived from this is that there are at least *three* types of sex chromosomes in some species of fish; the two traditional chromosomes, X and Y, plus an extra, W, chromosome. Visually, all three may be indistinguishable from each other but their distinct cytological existence can be determined through the inheritance of certain sex-linked characters.

Throwing the whole lot together – W, X and Y chromosomes, M and F factors, morphologically distinct sex chromosomes and cytologically distinct chromosomes – we end up with an overall picture of sex determination in fish which is extremely difficult to pin down. A further twist to the story is added by the 'power' or 'valence' of the M and F factors themselves. The actual locations of these appear not to be known with any degree of precision (other than that they are located both on autosomes and sex chromosomes). We do not, for example, know at the moment how many are located on X chromosomes or how many on Y.

What we do know is that their positioning within autosomes and sex chromosomes does not appear to be constant. Since sex determination seems to be controlled to a large extent by the relative potency of the M and F factors on the sex chromosomes and not by their numbers, it is possible to see how individual males and females can possess very different W, X and Y combinations, depending on the exact distribution of the various M (male) and F (female) determinants on the sex chromosomes.

For example, fish having the traditional sex distribution of XX=female, XY=male, include the Goldfish (*Carassius auratus*), the Siamese Fighter (*Betta splendens*) and the Medaka (*Oryzias latipes*). When it comes to the Platy (*Xiphophorus maculatus*), though, we can have, at least, WY, WX and XX females, and XY and YY males.

Despite all these difficulties, though, some form of general statement concerning sex determination can be made. Sex determination in fishes is polygenic in nature with M (male) and F (female) determinants located both on the autosomes and the sex chromosomes themselves (which may or may not be morphologically distinct from each other or from the autosomes). In some cases, three types of cytologically different sex chromosomes have been identified: W, X and Y. The Z chromosome, which is referred to in some scientific literature, is believed to be equivalent to the Y chromosome. Although the basic sexual potential that an individual fish possesses is genetic in nature, its ultimate manifestation may be influenced by some environmental factors.

Sex Determination in Poeciliines and Goodeines

The vast majority of fishes have very small chromosomes, which can make their detailed analysis very difficult indeed. Some types, however, among them the Goodeines, have largish chromosomes which considerably facilitate the search for morphological differences between males and females.

Despite this, very few chromosomal differences have been found. One outstanding exception is the Goodeid *Allodontichthys hubbsi*. In this species, the diploid number of chromosomes (referred to as 2n) has been found to be 42 in females and 41 in males. *A. hubbsi* therefore represents a very rare instance in fish in which an isolated cell can be identified as being male or female in origin. Since chromosomes occur in pairs, the 2n figure for most species is even in number. This being the case, the situation that exists in *A. hubbsi* is most unusual indeed. It can, however, be explained away quite simply – the 'missing' male chromosome is not really missing but has become fused to another, which is now identifiable as the Y chromosome.

As far as the other Goodeines are concerned, though, there seem to be no clear-cut differences between male and female cells. Genetic analyses have shown that the chromosome numbers vary betwen 24 and 48 in the 35 or so species studied, but that there are no detectable differences between the sexes at the cellular level.

Sexual determination has been more extensively studied among the Poeciliines, but even here, only a relatively small handful of species have been more or less comprehensively investigated.

Most Poeciliines have 48 chromosomes. Among these, are *Xiphophorus helleri*, the Swordtail, and *X. maculatus*, the Platy. The six *Poecilia* species studied, however, have 2n=46 as the norm, while the two species *Gambusia affinis* and *G. holbrooki* both have 2n=48, but have distinct sex chromosome differences. *G. affinis*, although it looks very similar to *G. holbrooki*, has a different arrangement of gonopodial ray modifications – it lacks the denticles that *G. holbrooki* has on ray 3, but has a longer segmented claw on 4p (the posterior branch of ray 4). In addition, wild populations of *G. holbrooki* produce melanic males from time to time, the melanism ranging from just a few spots to virtually totally black individuals. This characteristic has been traced to the Y chromosome. It is regarded as a holandric trait, found only in males and transmitted directly

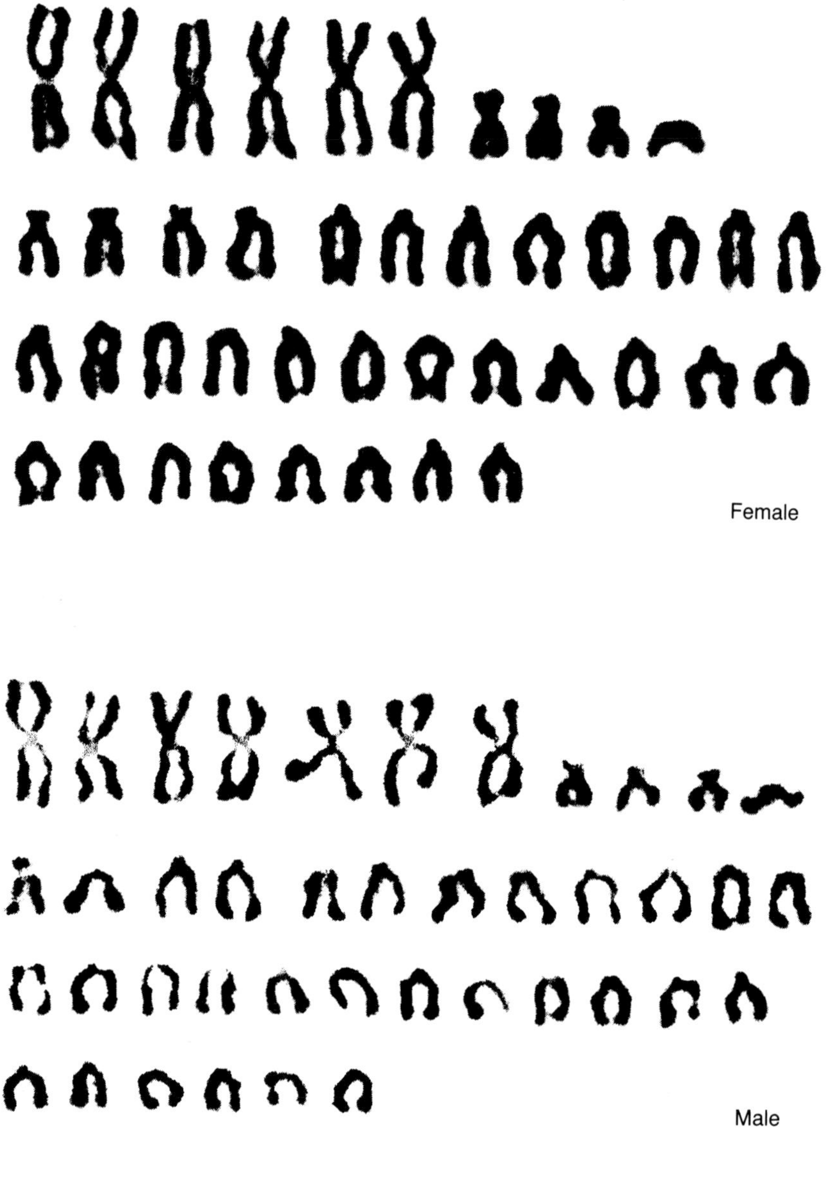

Female

Male

Fig. 16 *Allodontichthys hubbsi* is exceptional among fish in having distinct male and female sets of chromosomes. (Redrawn from T. Vyeno and R. R. Miller, 'Allodontichthys hubbsi, a new species of Goodeid fish from Southwestern Mexico', Occasional Papers of the Museum of Zoology, University of Michigan, No. 692, April 24, 1980)

from father to son. This can only happen if the gene responsible is not blocked in some way by an equivalent, but different, form of the gene on a corresponding sex chromosome. The fact that this does not occur indicates that male *G. holbrooki* are genetically XY males, the females being XX. In *G. affinis*, however, the sex determining factors behave according to a WZ/ZZ pattern (which is equivalent to WY/YY). What we have, therefore, are two very closely related fish whose sex determining system is quite different.

In Guppies (*Poecilia reticulata*), sex determination follows the standard XX/XY format, although discrepancies can occur when the M and F factors on the autosomes outweigh those on the sex chromosomes. In particular, atypical XX males can be produced when the M autosomal factors dominate the F factors on the X chromosomes.

In the Sailfin Mollies, *Poecilia velifera* and *P. latipinna*, experimental results indicate that they, too, have a more or less straightforward (XX=female, XY=male) genetic mechanism. *P. sphenops*, the shorter-finned Sphenops Molly, appears to have XX and XY males and WZ/ZZ (i.e. WY/YY) females.

Sex determination takes an unusual twist in the Amazon Molly, usually referred to as *Poecilia 'formosa'*. Known from a number of localities in Mexico, this 'species' looks very much like one of the Green Mollies and has differing characteristics, depending on the particular locality in which it is found. This, in itself, is no earth-shattering feature, of course. What makes the Amazon Molly special, though, is that only females are known.

When it was first described in 1859 by Girard, *P. 'formosa'* was deservedly hailed as a major novelty because it was the first all-female vertebrate ever discovered. Gradually, as the story has unfolded over the years, detailed aspects of the overall picture have shown it to be even more remarkable. It is quite a complicated issue, in which some of the finer points are still uncertain, making further revision, as ever more precise information comes to light, a strong possibility. In essence, the main points are as follows.

When *P. 'formosa'* was first discovered, it was given the common name of Amazon Molly because of its all-female composition (as in the legendary all-female race of Amazonian warriors). Its intermediate appearance, when compared to two other species of Molly with which it was known to co-exist in different parts of its range, was also noted. These two species, *P. latipinna* and *P. mexicana* (the latter regarded as *P. sphenops* at the time the early observations were made), were also known to co-exist in a number of localities. Further, where they did co-exist, *P. 'formosa'*-type hybrids were found. However, they differed from the standard *P. 'formosa'* in that both males and females of the hybrids occurred in these areas of overlap. Today, we know that all-female *P. 'formosa'* are found in areas where they co-exist either with *P. latipinna* or *P. mexicana*, but not where these two species are known to co-exist with each other.

This leads to some very interesting conclusions. In the first place, *P. 'formosa'* is believed to be a naturally occurring hybrid between *P. latipinna* and *P. mexicana*. It is also known to be a diploid having 46 chromosomes, the same number as both of these species. Further, *P. 'formosa'* needs to mate in order to reproduce. However, since there are no *P. 'formosa'* males, mating takes place between a *P. 'formosa'* female and a male from one or other of the two Molly species with which it co-exists.

In some areas, therefore, the sperm will come from a *P. latipinna* male, while in others the donor will be a *P. mexicana* male. Irrespective of the source, the sperm activate the process of cell division in the fertilised eggs but do not contribute any genetic material to the embryos. The mechanism whereby the sperm are prevented from making a chromosomal contribution is not fully understood, but, whatever the case, the result is that *P. 'formosa'* embryos receive all their genetic material from their mothers. It is this that results in all-female offspring. Sex determination in the Amazon Molly is therefore very much a question of male genetic information being omitted after fertilisation. This method of reproduction – which, although dependent on sperm to activate cell division, excludes all male chromosomal contribution – is known as *'gynogenesis'*.

A further twist to the *P. 'formosa'* story warrants mention here.

Laboratory experiments carried out by various workers over a number of years resulted in all-female offspring that were triploids – they had three times the haploid number of chromosomes. These triploid *P. 'formosa'* were produced as a result of a breakdown in the gynogenetic process outlined above, somehow resulting in the incorporation of male chromosomes in the overall genetic content of the embryonic cell, this giving them 69 chromosomes in total.

Triploids such as these are also known to exist in wild populations, and in significant numbers. Their genetic make-up depends, of course, on which of the two parental Molly species is involved. Some populations have triploids fathered by *P. latipinna* males, while others are linked to *P. mexicana*. The two types of triploid are referred to as *P. 2 latipinna–mexicana* and *P. 2 mexicana–latipinna* (some texts refer to them as *P. mexicana–2 latipinna* and *P. 2 mexicana–latipinna*, respectively).

Sex determination in these fish, as in the majority of others dealt with in this chapter, is detemined by the total potency or valence of female and male factors which, in these cases, are overwhelmingly female.

In summary, therefore, there appear to be three types of *P. 'formosa'* in the wild.

The first is a diploid (2n=46), arising through hybridisation between *P. latipinna* and *P. mexicana*, whose total genetic make-up is maternally derived and whose offspring therefore carry only female sex chromosomes. Reproduction is through gynogenesis. These fish can be referred to scientifically as viable *P. mexicana × latipinna* hybrids.

Then there are two triploids (3n=69), which may have arisen through a breakdown in gynogenesis that allowed incorporation of male chromosomes into the female cell nuclei. These have nevertheless been swamped by the surplus of female factors so that all the resulting fish are female. Scientifically, these triploids may be refered to as *P. 2 latipinna–mexicana* and *P. 2 mexicana–latipinna*, depending on parentage.

Since the discovery of all-female populations of the Amazon Molly, other superficially similar instances have come to light among the Poeci-liines. The genus in question is *Poeciliopsis*, and the all-female populations belong to three 'species'.

At first sight, the situation appears to be identical to that found in *P. 'formosa'*. However, detailed experimental analysis of inherited characters carried out by Schultz and others shows a clear paternal influence in the offspring (in gynogenesis as exhibited by *P. 'formosa'*, the paternal chromosomes are somehow inactivated). Therefore, the mechanism is not quite the same in these *Poeciliopsis* 'species'. The populations may be all-female, but the inheritance of paternal characters indicates that the males involved in the crosses do contribute something, even though the female genes are sufficiently strong to override any male influences as far as the sex of the offspring are concerned.

A long series of 'shotgun' experiments in which all manner of crosses were attempted, followed by a series of highly sophisticated investigations, has yielded some very interesting results. It was discovered, for example, that one all-female 'species' is a hybrid between *Poeciliopsis monacha* and *P. lucida*; that another is a hybrid between *P. monacha* and *P. occidentalis*; that a third, unisexual, 'species' (from the Río Fuerte) appears to be a hybrid between *P. monacha* and *P. latidens*; and that a similar hybrid from the Río Mocorito also appears to be a *P. monacha × latidens* hybrid, but there are indications of *P. viriosa* genes as well which, if so, would make this fish a *trihybrid*, a hybrid involving three parental species.

The first two of these 'species' appear to have been better studied than the others, especially in terms of paternal–maternal genetic relationships, where it has been found that the obvious paternal influence in the first generation becomes somehow fixed in the all-female offspring. Once these females mature, they produce eggs by the process of reduction-division (meiosis). At this stage, only the maternal genes appear to survive, the male ones being lost in the process. Once the eggs have been fertilised and have received their dose of paternal genes, the whole cycle starts again. In this way, we end up with an all-female fish whose phenotype, or physical appearance, remains constant. This process of paternal contribution followed by the subsequent loss of paternal chromosomes at meiosis is known as *hybridogenesis*.

Of the many species comprising the genus *Xiphophorus*, the one that has been most extensively studied in terms of sex-determining factors is *X. maculatus*, the Platy, where three sex chromosomes – W, X, and Y – have been identified. The W chromosome, matched either to an X or a Y, is found only in females. Paired X chromosomes are also restricted to females, while males have various X and Y combinations. Genetically, females can therefore be WY, WX or XX, while males are either XY or YY.

It has also been shown that, in this species, fish possessing a Y chromosome develop, under normal circumstances, into males. This would indicate that WY females are females because there is some factor on the W chromosome (probably a suppressor gene) that somehow blocks the influence of the male factor in the Y chromosome.

Under experimental conditions, though, WY, WX and XX individuals can all develop into males even though, genetically, they appear to have female configurations. This implies that male factors are present in *all* the sex chromosomes, but that some form of mechanism prevents them from being expressed on the W and X chromosomes under normal circumstances. Klaus Kallman (1984) postulated quite a simple way in which this could work. All that is needed is the existence of one or more genes that block the activation of male genes on the W and X chromosomes and the suppressor gene on the W that will block the male gene (or genes) on the Y chromosome in the case of WY females.

Despite the close relatonship between *X. maculatus* and *X. helleri* (the Swordtail), *X. milleri* (the Catemaco Livebearer), *X. montezumae* (the Montezuma Swordtail), *X. nigrensis* (one of the Dwarf Helleris/Pigmy Swords), and others, sex-determination analysis in these other species has not proceeded quite at the same pace. They are all believed to exhibit X/Y sex chromosomes with the standard XX=female, XY=male arrangements (though deviations such as XY females are occasionally encountered in, for example, *X. montezumae* and *X. milleri*, and XX males in *X. nigrensis*). Further, small early-developing and large late-developing *X. helleri* males, along with an over-abundance of females in late-maturing populations, indicate that, while X and Y chromosomes appear not to have been unequivocally identified in this species, they, along with autosomal sex-determining factors, do, nevertheless, exist.

As the above selection of examples demonstrates, sex determination in the Poeciliines does not always follow the same basic pattern. If we then superimpose the possible effects of environmental factors on the genetic ones, the situation becomes even more interesting.

For example, a well-documented experiment regarding the effects of temperature on sex determination in *Poeciliopsis lucida* shows beautifully how environment and genetics interact.

Sullivan and Schultz (1986) reared genetically controlled strains of *Poeciliopsis lucida* under four different temperature regimes and obtained the following results with one of the strains:

Temperature (°C)	% Males in Population
24.0	38
25.5	54
27.0	63
30.0	92

There can be little doubt, therefore, that temperature *does* play a key role in sex determination, at least in this species. However, results from a second strain showed no significant deviation from the expected 1 male:1 female ratio, irrespective of the temperature.

What this proves is that, while environmental factors (in this case, temperature), can affect sex determination in *P. lucida*, the genetic potential to respond to these changes has to be there in the first place. In other words, sex in *P. lucida* is under genetic control, but it can be influenced by external factors.

In *Xiphophorus helleri*, pH was demonstrated to have a profound effect on sex determination. Rubin reported, in 1985, that broods reared at pH 6.2 showed 100% differentiation into males, while those reared at pH 7.8 showed 98.5% differentiation into females.

While these results appear pretty conclusive, they have been criticised on several grounds. For instance, they have been neither confirmed nor repeated in any Poeciliine, not even in *X. helleri* itself. Other criticisms include lack of control of the genetic background of the fish used and the small size of the experimental samples (two broods at each pH level). Therefore, until similar, well-controlled and meticulously documented experiments are carried out, Rubin's results may not receive universal acceptance.

If and when the experiments are repeated, what we are likely to find is that pH, like temperature, has a profound effect on sex determination in Poeciliine livebearers. Again, as in the *P. lucida* temperature-related example cited above, it seems reasonable to assume that reaction to pH changes may show varying degrees of intensity depending, not only on the pH values themselves, but also on the varying genetic potential of different strains to react to a changing chemical environment.

Conclusion

Most fish, like other vertebrates, have fixed, distinct, separate sexes. Some, however, can change sex under environmental influences, often brought about by the presence or absence of a dominant individual. Prominent among these sequential hermaphrodites are Wrasses, which pass from juvenile to female and finally to male (protogynous hermaphroditism), and Clownfish, which pass from the juvenile phase through a male phase, ending up as females (protandrous hermaphroditism). Other fish, like some of the Sea Perches, can be synchronous, or functional, hermaphrodites, in which an individual can exhibit both sexes simultaneously.

Sex reversal from female to male – that is, of the protogynous type – has been reported in Poeciliine livebearers, particularly the Swordtail, *Xiphophorus helleri*. Despite the well-publicised nature of these reports, unequivocal scientific proof is still lacking.

In general terms, sex determination in fish is polygenic or polyfactorial – that is, it is controlled by many genes located both on the sex chromosomes and on the autosomes.

In addition to the sex chromosomes, which may or may not be distinguishable, either from each other or from the autosomes, fish possess M (male) and F (female) determinants, the potency or valency of which plays a vital role in determining the sex of an individual. In at least one species, the Goodeid *Allodontichthys hubbsi*, males have one chromosome fewer than females, making them a bit of an exception in that the sex of an individual fish can be determined from an examination of an individual cell.

Three types of sex chromosome are known to exist in some species of fish, such as the Platy (*Xiphophorus maculatus*). These are the normal X and Y chromosomes, plus an additional one referred to as the W chromosome. Reference may also be found in some literature to a Z chromosome, but this is now generally accepted as being equivalent to the Y chromosome.

Precise knowledge of the locations of the M and F determinants on the sex chromosomes and autosomes is not currently available, but it is known that their distribution is a variable factor. This leads to a situation where both males and females can have a range of genotypes (genetic make-ups).

A further influential factor appears to be the type of environment in which a developing brood of fry finds itself. Temperature, for example, has a profound effect on sex determination in at least one species, *Poeciliopsis lucida*, but pH may also play a role.

In unisexual 'species', such as the naturally occurring Amazon Molly (*Poecilia 'formosa'*) and several *Poeciliopsis* hybrids, sex is determined by the overwhelming influences of maternal chromosomes or the total lack of male genes. Male sperm are used either to initiate cell division in an egg, with the subsequent inactivation of the male genes (as in gynogenesis in *P. 'formosa'*), or to fertilise an egg in the real sense of the word, with the ensuing incorporation of paternal traits, followed by the eventual loss of the male genetic material during the production of eggs in the mature fish (this is known as hybridogenesis and is exhibited by *Poeciliopsis* hybrids).

Further Reading

Kallman, Klaus D., 'A New Look at Sex Determination in Poeciliid Fishes' in *Evolutionary Genetics of Fishes,* Bruce J. Turner (ed.), Plenum Press, p. 636 (1984).

Miller, Robert Rush and Uyeno, Teruya, '*Allodontichthys hubbsi*, a New Species of Goodeid Fish From Southwestern Mexico', *Occasional Papers of the Museum of Zoology, University of Michigan*, 692, pp. 1–13 (1980).

Monaco, Paul J., Rasch, Ellen M. and Balsamo, Joseph S., 'Apomictic Reproduction in the Amazon Molly, *Poecilia formosa*, and its Triploid Hybrids' in *Evolutionary Genetics of Fishes,* Bruce J. Turner (ed.), Plenum Press, p. 636 (1984).

Moore, William S., 'Evolutionary Ecology of Unisexual Fishes' in *Evolutionary Genetics of Fishes,* Bruce J. Turner (ed.), Plenum Press, p. 636 (1984).

Rubin, D. A., 'Effect of pH on Sex Ratio in Cichlids and a Poeciliid (Teleostei)', *Copeia*, 233–5 (1985).

Schultz, R. Jack, 'Origins and Relationships of Unisexual Poeciliids' in *Ecology and Evolution of Livebearing Fishes,* Gary K. Meffe, Franklin F. Snelson, Jnr., *et al.,* Simon & Schuster, p. 450 (1989).

Snelson, Franklin F., Jnr., 'Social and Environmental Control of Life History Traits in Poeciliid Fishes' in *Ecology and Evolution of Livebearing Fishes,* Gary K. Meffe, Franklin F. Snelson, Jnr., *et al.,* Simon & Schuster, p. 450 (1989).

Sullivan, J. A. and Schultz, R. J., 'Genetic and Environmental Basis of Variable Sex Ratios in Laboratory Strains of *Poeciliopsis lucida*', *Evolution* 40, 152–8 (1986).

Wooten, M. C., Scribner, K. T. and Smith, M. H., 'Genetic Variability and Systematics of *Gambusia* in the Southeastern United States', *Copeia*, 283–9 (1988).

Uyeno, Teruya, Miller, Robert Rush and Fitzsimons, John Michael, 'Karyology of the Cyprinodontoid Fishes of the Mexican Family Goodeidae', *Copeia*, 497–510 (1983).

Yamamoto, Toki-O, 'Sex Differentiation' in *Fish Physiology III*, Hoar and Randall (eds.), Academic Press (1969).

Distribution

One hundred years ago, it would have been possible to say that livebearers were found in tropical and sub-tropical regions of the New World. Today, the picture is quite different, at least in the case of the Poeciliid livebearers, many species of which have been intentionally or accidentally introduced into locations outside their natural range. Goodeid livebearing species, Halfbeaks, *Corynopoma*, *Oryzias* and the two Killies that exhibit facultative viviparity have not, however, been subjected to the same treatment.

Poeciliinae

Nineteen Poeciliine species are known to be found as wild, reproducing, and therefore more-or-less sustainable, populations in locations lying outside their original ranges. So, when we speak about the distribution of Poeciliines, perhaps we should now distinguish between their natural and their non-natural distributions.

Of the introduced species, the most widespread of all is, undoubtedly, the Mosquito Fish, *Gambusia affinis*, which is now found in every continent except Antarctica and which may well be the most widely distributed fish in the world.

As its common name indicates, this species has been repeatedly introduced to control mosquitoes and, thus, the spread of malaria. It is known to have an almost insatiable appetite for mosquito larvae and pupae. Unfortunately, it has the same appetite for *anything* that is small enough to be swallowed, including young fish. This means that its introduction into any area soon results in a fast increase in its numbers, accompanied by a corresponding decrease in the native population. The problem is compounded by the fact that *G. affinis* is very aggressive; it will nip the fins of larger fish, often to the point where they are so weakened and harassed that they succumb to infection and eventually die. Therefore, while the mosquito-controlling abilities of this species may be considerable, its overall impact must be regarded as negative, especially since most fish species affected by the introductions would, in any case, themselves feed on mosquito larvae and pupae.

Reports of *G. affinis* introductions abound, ranging from Canada, parts of the United States outside the species' natural distribution (Mississippi River Basin to Iowa, coastal drainages of the Gulf of Mexico, down to Veracruz in Mexico and the Atlantic slope up to New Jersey), to Japan,

Egypt and most of Europe. In addition, I have collected this species in locations as widely separated from each other as Madeira, Gibraltar, southwestern Spain, Singapore and Malaysia.

G. affinis is a species much maligned by ichthyologists and aquarists because it is a super-efficient destroyer of native species. While not disputing this in any way whatsoever, I nevertheless must confess to having a great deal of affection for this species, most probably because it was the first fish I ever kept and bred, way back in 1952.

It is ironic that such a robust species as *G. affinis* belongs to the same genus as some of the most endangered of all the livebearers, such as *Gambusia gaegei*, *G. georgei* and *G. heterochir*, all of which have been introduced into controlled localities outside their natural (southwestern United States) range almost as a last-ditch conservation measure. The same applies to the highly endangered *Poeciliopsis occidentalis*.

Despite Courtenay's and Meffe's exhaustive research, which I have largely relied on in Table IX, plus my own observations and those of several ichthyologists, there can be little doubt that the actual distribution of Poeciliid livebearers outside their natural range is even wider than indicated here and is likely to continue to expand.

Some introductions are easy to understand, like those associated with mosquito control. Others occur within the same broad geographical region naturally occupied by the species in question, as in the case of Mexican introductions of the Sailfin Molly (*Poecilia latipinna*). These are not generally regarded as 'unexpected' or 'surprising' even though we may deem them regrettable.

Some, though, are quite unfathomable. Why, for instance, would anybody wish to release *Phalloceros caudimaculatus*, which is a South American species, into Malawian waters? The same could be asked about the Argentinian species *Cnesterodon decemmaculatus*, which is now found in Chile. Neither of these two species, delightful though they undoubtedly are in their own ways, could be regarded either as highly popular or widely available, thus making their establishment in exotic localities a bit of a puzzle.

TABLE IX   DISTRIBUTION OF POECILIINES WITH EXOTIC
POPULATIONS (LARGELY BASED ON COURTENAY & MEFFE, 1989)

SPECIES

Scientific Name   *Belonesox belizanus*

Common Names   Pike Top Livebearer; Pike Killifish

DISTRIBUTION

Natural
Atlantic slope of Middle America (from Laguna San Julian, Mexico to Costa Rica)

Non-natural
Southwestern Dade County and North Key Largo in Florida; probably San
Antonio River in Texas

SPECIES

Scientific Name   *Cnesterodon decemmaculatus*

Common Name   None

DISTRIBUTION

Natural
Argentina

Non-natural
Chile

SPECIES

Scientific Names   *Gambusia affinis* and *G. holbrooki* (No distinction is made here
between the two species)

Common Name   Mosquito Fish

DISTRIBUTION

Natural
Mississippi River drainage south of southeast Iowa; coastal regions of Gulf of
Mexico down to Veracruz; Atlantic slope of US up to southern New Jersey

Non-natural
Arizona; California; Hawaii; Montana; Nevada; New Mexico; Oregon; Utah;
Washington; Wyoming; Alberta; possibly British Columbia and Manitoba;
Sonora; Baja California Sur; Chihuahua; Puerto Rico; Argentina; Bolivia; Chile;
Peru; Europe, including Spain, Tenerife, Italy, Gibraltar, Portugal, Madeira,
Hungary and Yugoslavia; Central African Republic; Libya; Egypt; Israel;
Madagascar; South Africa; Sudan; Zimbabwe; Ghana; Ivory Coast; Annobon

Island; Gulf of Guinea; Singapore; Malaysia; American Samoa; Western Samoa; Australia; Cook Islands; Federated States of Micronesia; Fiji; Guam; Kiribati; New Zealand; Mariana Islands; Papua New Guinea; Tahiti (There are, almost certainly, numerous other localities that have not been officially recorded.)

SPECIES

Scientific Name   *Gambusia panuco*

Common Name   Panuco Gambusia

DISTRIBUTION

Natural
Río Panuco in Mexico

Non-natural
One known locality in Mexico: Media Luna Springs in San Luis Potosí – an area famous for the existence of the Blind Cave Characin (*Astyanax fasciatus mexicanus*, formerly *Anoptichthys jordani*)

SPECIES

Scientific Name   *Limia vittata* (formerly *Poecilia vittata*)

Common Name   Cuban Limia

DISTRIBUTION

Natural
Cuba and Isle of Pines

Non-natural
Hawaii (Oahu)

SPECIES

Scientific Name   *Phalloceros caudimaculatus*

Common Names   One-spot Livebearer; Caudo

DISTRIBUTION

Natural
Southern Brazil from around Río de Janeiro, Paraguay, Uruguay and Argentina (between latitudes 20° and 35°S)

Non-natural
Unlikely though it may seem: Malawi and Western Australia

SPECIES

Scientific Name   *Poecilia 'formosa'*

Common Name   Amazon Molly

DISTRIBUTION

Natural
Lower Rio Grande Valley (Texas) to Río Tuxpan estuary (Mexico)

Non-natural
Rio San Marcos and Rio San Antonio, both in Texas

SPECIES

Scientific Name   *Poecilia latipinna*

Common Name   Sailfin Molly

DISTRIBUTION

Natural
Gulf of Mexico from Yucatán northwards and eastwards into Florida and North Carolina

Non-natural
Arizona, California, Nevada, Texas, Alberta (Canada), states of Hidalgo and Sonora in Mexico, Central America, Singapore, Australia, New Zealand, Philippines, Guam and Hawaii. Some Black Mollies, probably from *P. latipinna* × *P. velifera* crosses, are reportedly found in a number of southern states of North America. A similar black population is also established in an overflow stream at Al Qatif oasis in the Eastern Province of Saudi Arabia. A neighbouring overflow

Plate 22   Black Mollies are known from a few localities outside their normal range, including an oasis in Saudi Arabia. (*Florida Tropical Fish Farms Association*)

stream at the same oasis has variously coloured *P. latipinna*, including wild-type (green), marbled and 'poor-quality blacks that are smaller' and have a faint orange margin on the dorsal fin. Although the exact source of these fish is not known, it is thought likely that they have come from the collections owned by the large expatriate and local community of aquarists

SPECIES

Scientific Name *Poecilia mexicana* (and the allied species, *P. sphenops* and *P. butleri*)

Common Names Short-finned Molly or Atlantic Molly (*P. mexicana*); Green or Sphenops Molly (*P. sphenops*); Pacific Molly (*P. butleri*)

DISTRIBUTION

Natural (for *P. mexicana*)
Atlantic slope of Middle America (Rio San Juan in the Grande basin), Nuevo León in Mexico, Pacific slope (Río del Fuerte Basin), Sonora in Mexico, southwards to Colombia (Caribbean slope), Pacific slope (Río Tuira) in Panama and Dutch and Colombian West Indies

Non-natural
California (north of the Rio Grande basin), Idaho (Bruneau River); Nevada (springs and Moapa River); possibly Colorado (Saguache.County) and parts of Florida. American Samoa, Western Samoa, Fiji, Tahiti and Hawaii have also been reported as having established populations. The Green Molly (*P. sphenops*), and *P. butleri*, both closely related to *P. mexicana*, have been reported from Oahu in Hawaii (*P. sphenops*) and 'several islands of Oceania' (possibly *P. butleri*). Definitive identification of both these species is still awaited

SPECIES

Scientific Name *Poecilia reticulata*

Common Names Guppy; Millions Fish

DISTRIBUTION

Natural
Not as straightforward as for most of the others. Definite natural range includes the Dutch Antilles, Trinidad, Barbados, St Thomas and Antigua (both of these belonging to the Leeward Islands), Venezuelan islands and western Venezuela through to Guyana. Possible natural populations include those of northeastern Venezuela, Margarita, Tobago, and Lesser Antilles

Non-natural
After *Gambusia affinis*, *P. reticulata* is the most widespread introduced Poeciliine species. Populations are known to exist in Arizona, California, Florida, Idaho, Nevada, Texas, Wyoming, Alberta (Canada) and Mexico, though there is some

doubt as to the self-perpetuating nature of some of these. Reports of other introductions, most of which now consist of established populations, include Nigeria (the males are mostly pinkish/yellowish), Kenya, Uganda, Colombia (multi-coloured but small males), Peru, Puerto Rico, Australia, New Zealand, Fiji, Guam, Hawaii, Palau, Tahiti and Western Samoa. Populations in South Africa and in thermal waste water in the United Kingdom (near St Helens) have both disappeared. Most Guppy introductions, as in *Gambusia affinis*, have been intentional ones to combat malarial mosquitos, though escaped and released aquarium stocks no doubt also feature quite prominently, as in the reported case of a few 'stray' specimens found in the overflow streams of the Al Qatif oasis in the Eastern Province of Saudi Arabia. Whether or not these last have developed into a self-sustaining population is not known at the time of writing.

SPECIES

Scientific Name   *Poeciliopsis gracilis*

Common Name   Porthole Livebearer

DISTRIBUTION

Natural
Atlantic and Pacific slopes of Middle America from southern Mexico to Honduras

Non-natural
Drainage ditch near Mecca (California) and Venezuela. The former population may no longer be extant.

SPECIES

Scientific Name   *Xiphophorus helleri*

Common Name   Swordtail

DISTRIBUTION

Natural
Atlantic slope of Middle America from Veracruz in Mexico (Río Nautla) to northwestern Honduras (between latitudes 12° and 26°N)

Non-natural
X. helleri can be found outside its natural range, either in the 'pure' state or as hybrids (usually with *X. maculatus*, the Platy). The actual status (established or otherwise) of all the introduced populations is not known but there are reports of introductions ranging from Florida to Indian Spring, and upstream of the Overton Arm of Lake Mead (both in Nevada), Kelly Warm Spring (Wyoming), Arizona, California, thermal waters in Alberta (Canada), Hawaii, Puerto Rico, Michoacán, Morelo, Nuevo León and Coahuila (these last four localities are in Mexico), South Africa (in Transvaal), Sri Lanka, Australia, Guam, Sulawesi and Fiji. A population at St Helens (UK) disappeared sometime in the mid-1980s

Plate 23 The Porthole Livebearer (*Poeciliopsis gracilis*) has been recorded from Venezuela and California, both being locations outside its normal territory. (*Dennis Barrett*)

SPECIES

Scientific Name   *Xiphophorus maculatus*

Common Names   Platy or Southern Platyfish

DISTRIBUTION

Natural
Atlantic slope of Middle America from south of Ciudad Veracruz to northern Belize

Non-natural
As in the case of the Swordtail, the Platy occurs outside its natural range, both in the pure form and as hybrids with *X. helleri*. It is now found, with varying degrees of permanence, in thermal water in Nevada (one locality, a spring), Florida, Orange County in California, Texas and Nigeria. Definitely established populations exist in Hawaii (three islands), Coahuila, Guanajato, Nuevo León and Sonora (all in Mexico), Puerto Rico and Australia. Other possible localities are Kiribati (in Palau) and Colombia, while the odd specimen has been reported from two overflow streams in the Al Qatif oasis in the Eastern Province of Saudi Arabia. The most significant introduction/release must be the one that took place prior to 1976 into the Río Teuchitlán, Mexico, which resulted in the extinction of the endemic Goodeid *Skiffia francesae*

SPECIES

Scientific Name   *Xiphophorus variatus*

Common Names   Sunset, Variatus or Variable Platy

DISTRIBUTION

Natural

This is a strictly Mexican fish found from southern Tamaulipas to eastern San Luis Potosí and northern Veracruz

Non-natural

Found, with varying degrees of permanence, in Alachua County and other counties in Florida, thermal springs in Montana (three counties), Salt River in Tempe, Arizona (this population was, reportedly, wiped out in 1965), Tuma (also in Arizona), Orange County and Riverside County in California, Colombia and Oahu (Hawaii)

Goodeinae

Goodeids are strictly Mexican (hence one of their common names: Mexican Livebearers) fish whose distribution is centred around the Mesa Central with a concentration of species in the Río Lerma basin. Most species are found in upland waters, a feature that has given rise to a second, little-used common name – Highland Carp.

The most widespread genus is *Ilyodon* (the main exception to the 'highland' tag), which occurs from an area southeast of Mexico City (Morelos) westwards and slightly northwestwards, past Guadalajara (in Jalisco) and Colima, reaching almost to the Pacific coast of Mexico.

The area around Durango is characterised by the genus *Characodon*. *Goodea* is found around San Luis Potosí, about halfway between the Gulf of Mexico in the east and the Pacific Ocean in the west, extending westwards almost to the Pacific coast and southwards towards the region west of Mexico City.

Xenoophorus occurs mainly around San Luis Potosí, while *Ataeniobius* is found slightly further eastwards. *Xenotoca* is predominantly a west-Mexican genus, occurring in a roughly broad band whose easternmost boundary can be traced by an imaginary north–south line extending from San Luis Potosí to an area roughly between Mexico City and Guadalajara, taking in Michoacán and Jalisco. *Allotoca* is also found within this same area, although it does not extend quite as far westward.

Skiffia is found west and southeast of Guadalajara, mainly in the state of Jalisco, but one species, *S. lermae*, occurs in Michoacán. *S. francesae*, as already mentioned, is no longer found in its original habitat in the Río Teuchitlán in Jalisco, the only surviving specimens known to exist being

Plate 24   Goodeids (this is *Ameca splendens*) are strictly Mexican. (*Florida Tropical Fish Farms Association*)

found in the aquaria of specialist hobbyists in the United States, the United Kingdom and West Germany.

Hubbsina is predominantly a mid-Mexican genus. It has its focus in an area of Michoacán that lies halfway between Mexico City in the east and Guadalajara in the west. *Girardinichthys* is centred largely upon Mexico City, the most famous locality being Chapultepec Park in the City itself, where the first Amarillos, *Girardinichthys viviparus* (known as *Limnurgus innominatus* at the time), were collected. However, attempts to collect this species from the now highly polluted lake in Chapultepec Park in 1988 proved unsuccessful and there must now be serious doubt regarding the continued existence of Amarillos at this locality.

Zoogoneticus, like *Xenotoca*, is largely a western fish, its type locality being Laguna Chapala (southeast of Guadalajara) but extending eastwards towards Michoacán and westwards towards the Pacific coast. *Allodontichthys* is a southwestern genus found in Michoacán, Jalisco and Colima to the southwest of Guadalajara. This is also the rough area in which *Xenotaenia* is found.

The greatest concentration of Goodeine species occurs in an area just south of Guadalajara (usually referred to as the Río Lerma basin) and its associated basins such as the Río Ameca basin, home of *Ameca splendens*, perhaps the best-known of all the Goodeids.

This central zone of abundance contains many of the genera also found elsewhere, like *Allodontichthys*, *Alloophorus*, *Allotoca*, *Ameca*, *Chapalichthys*, *Goodea*, *Ilyodon*, *Neoophorus*, *Skiffia*, *Xenotoca* and *Zoogoneticus*.

Anablepinae

There are only two genera of Anablepids that fall within the scope of this book – *Anableps* (incorporating the various Four-eyed Fishes) and *Jenynsia* (the One-sided Livebearer). *Anableps* is found in fresh waters and brackish coastal waters of Central America, both the Atlantic and Pacific slopes, and northern South America. *Jenynsia* occurs considerably further south in the lowlands of Brazil, Paraguay, Uruguay and Argentina. Kurt Jacobs (1971) described its distribution as 'south of the Amazon, between latitudes 20° and 35°S'.

Hemirhamphinae

The Halfbeaks featured in this book all have a Far-Eastern distribution pattern. The main centres for *Dermogenys*, the best-known of the three genera, are Malaysia, Sumatra, Java, Thailand, Kalimantan (Borneo), Sulawesi (Celebes) and the Philippines. *Hemirhamphodon* is centred upon Kalimantan, Malaysia, Singapore and Sumatra and *Nomorhamphus*, the most restricted genus in terms of distribution, upon Sulawesi.

None of the Halfbeaks have been the subject of translocations into non-native localities outside their normal ranges.

Rivulidae

This large family of Killifishes, which includes the genus *Cynolebias*, has a New World distribution pattern, ranging from south Florida to the Bahamas, Cuba, Hispaniola, Trinidad, Middle America down through South America as far south as Uruguay.

The genus *Cynolebias* is found in parts of Venezuela and the lowlands of Argentina, Brazil and Paraguay. Its two species, *C. brucei* and *C. melanotaenia*, are Brazilian in origin. Neither is found in any large numbers within the hobby, and neither, to the best of my knowledge, has been introduced into exotic locations.

Characidae (Géry) ≡ Characinae (Nelson)

The Characins are strictly freshwater tropical fish found in southern North America, Central and South America and Africa.

The tribe Glandulocaudini (the Croaking Tetras) contains the species *Corynopoma riisei*, the Swordtail Characin, which is distinguished from all other characoids in that it exhibits facultative viviparity. The tribe as a whole is widely distributed in tropical South America, while *Corynopoma* itself is found mostly in the upper Río Meta in Colombia and Trinidad.

Oryziidae/Oryziatidae (Nelson) ≡ Adrianichthyidae (Rosen & Parenti)

The Medakas or Rice Fishes are fresh- and brackish-water fish widely distributed from India and Japan to the Indo-Australian Archipelago. The only species of interest to us is *Oryzias latipes* – variously known as the Geisha Girl Fish, Japanese Medaka, Golden Medaka or, simply, Rice Fish – a native of the paddyfields of lowland Japan.

Conclusion

The vast majority of livebearers are still found in their original habitats today, just as they were when they were first discovered. Two species – *Skiffia francesae* and *Gambusia amistadensis* – are now known or believed to be extinct in the wild. Of the others, fewer than twenty species (all Poeciliid livebearers) have been introduced into localities that lie outside their natural ranges. Few of these have undergone dramatic population explosions, but those few – notably *Gambusia affinis*, one of the so-called Mosquito Fishes introduced for biological anti-malaria control purposes – have altered the balance of the native populations of fish and, perhaps, other organisms, usually with significant, and negative, consequences.

Once an introduced species becomes well established and known within a host country, it soon becomes regarded as a naturally occurring species indigenous to the country in question. The world is full of such examples, from the ubiquitous rabbit, to goats, rats, squirrels and numerous other one-time exotics. The problem with this is that people begin to lose sight of the true nature of their national fauna and flora – so much so that many Europeans, for example, think that *Gambusia affinis* is actually a European fish. In Singapore, *Oreochromis mossambicus*, the Mozambique Mouthbrooder, is sometimes referred to as the Japanese Fish, a name that indicates one of the sources of, at least, some of the introduced populations – but this species is no more Japanese than a Guppy is Nigerian!

As the foregoing pages will, hopefully, have shown, the global distribution of Poeciliid livebearers today is far from straightforward, and certainly much wider than it was, say, at the beginning of the century. It is also likely to become even more complicated with the passage of time.

The reproductive abilities of Poeciliid females are such that a single inseminated specimen can act as the source of a major 'infestation'. Aquarists and all others involved in the production and keeping of livebearers should therefore be especially diligent in preventing the escape of exotics into natural waters.

Such action can only add to an already regrettable distribution map which has seen the demise of numerous native species as a direct result of the sometimes well-intentioned, sometimes accidental, action of those who, perhaps, should have known better. On top of everything else, the release of exotics is actually illegal in many countries, so the unwary, or uncaring, or uninformed could end up not just disrupting the ecological balance of a particular habitat but breaking the law as well.

Further Reading

Courtenay, Walter R. and Meffe, Gary K., 'Small Fishes in Strange Places: A Review of Introduced Poeciliids' in *Ecology and Evolution of Livebearing Fishes*, Gary K. Meffe, Franklin F. Snelson, Jnr., *et al.*, Simon & Schuster p. 450 (1989).

Jacobs, Kurt, *Livebearing Aquarium Fishes*, Tropical Fish Hobbyist Publications, Inc. p. 495 (1971).

Meyer, Manfred K., Wischnath, Lothar and Foerster, Wolfgang, *Lebendgebärende Zierfische – Arten der Welt*, Mergus-Verlag p. 496 (1985).

Parenti, Lynne R., 'A Phylogenetic and Biogeographic Analysis of Cyprinodontiform Fishes (Teleostei, Atherinomorpha)', *Bulletin of the American Museum of Natural History*, Vol. 168, Article 4 (1981).

Ross, William, *Expatriated Mollies*, Tropical Fish Hobbyist Publications, Inc., pp. 59–63 (July 1984).

Uyeno, Teruya, Miller, Robert Rush and Fitzsimons, John Michael, 'Karyology of the Cyprinodontoid Fishes of the Mexican Family Goodeidae', *Copeia*, 2, pp. 497–510 (1983).

Part II
Aquarium Care

Aquarium Layouts

Guppies, Mollies, Swordtails and Platies of every shape and colour imaginable are sold in countless numbers worldwide, almost always as companion species to other community species, such as Neons (*Paracheirodon innesi*), Cardinals (*P. axelrodi*), Dwarf Gouramis (*Colisa lalia*), Corydoras Catfish (*Corydoras* spp), Angels (*Pterophyllum scalare*), Sucking Loaches or Chinese Algae Eaters (*Gyrinocheilus aymonieri*), Tiger Barbs (*Barbus tetrazona*), Zebra Danios (*Brachydanio rerio*) and the like.

What is often overlooked is the fact that this long-standing tradition does not necessarily mean that these, or other, livebearers are ideally suited for community aquaria.

Guppies

Guppies, famed for their hardiness, often survive in environments that would kill most other fish. This must not, however, be taken as a licence to abuse this delightful fish. Surviving is one thing, thriving is another.

Today's Guppies are all fancy varieties with long flowing fins. While this may make them attractive to hobbyists, there is no denying the fact that abnormally developed finnage slows a fish's movements down considerably.

Flowing fins and slow body movements seem to have an almost magnetic effect on fin-nipping species like Tiger Barbs. What often happens, therefore, is that Guppies are mercilessly harassed by the fast-swimming hyperactive Tiger Barbs, often to the point where their fins become badly nipped and their health badly affected as a result of stress. In the end, the weakened fish will succumb to some infection or other and die. Or they will spend most of their time hiding in a corner, as far away as possible from the disruptive influences of their tankmates.

There are two ways of alleviating such a situation. Firstly, it is known that Tiger Barbs and other fin-nipping species engage in this type of behaviour more frequently, and with greater intensity, when they are kept either as single specimens or as pairs. When, say, six or more individuals are kept together in a shoal, they seem to become so involved in minor internal squabbles that they tend to ignore most other inhabitants of the aquarium. So, what may at first appear foolhardy – increasing the number of potential fin-nippers – can in fact lead to a more relaxed life for previously over-stressed tankmates.

Secondly, the provision of ample cover in the form of dense clumps of vegetation, both rooted and floating, will allow Guppies and other harassed individuals to find shelter. Tiger Barbs, for their part, are predominantly clear-water swimmers, preferring to remain in shoals in the more open reaches of the aquarium. An adequately planted aquarium has the added bonus of looking attractive, of course.

Guppies can also be victimised by Angel Fish. While these attractive cichlids may not be excessively aggressive, they cannot really be regarded as angelic in their behaviour. They are, after all, predators, so anything that is small enough to be swallowed will be, quite naturally, regarded as a potential meal. Therefore, small Guppies, or Neons, or other small fish, and large Angels don't mix.

Swordtails and Platies

Swordtails and Platies generally fare better than Guppies in the company of hyperactive or potentially predatory species. Their larger size no doubt plays a part in this, since the 'I am not swallowable' message conveyed by the body dimensions of such a fish will not elicit the same predatory

Plate 25   Long-finned varieties of Sword, like this Painted Lyretail, often attract the attention of fin-nipping tankmates. (*Harry Grier/Florida Tropical Fish Farms Association*)

reactions that a small individual invariably does. Even so, long-finned varieties, such as the Lyretail Swordtails or Hi-fin Platies, can sometimes attract the attentions of fin-nippers just as Guppies do. In the case of Swordtails, their reluctance to succumb to intimidation may also be a contributing factor. Male Swordtails can be so aggressive towards each other that they are usually kept singly, or in shoals where their aggression is divided rather than focused on a single subordinate individual.

Mollies

When it comes to Mollies, there is an added dimension that needs to be borne in mind. While Swordtails, Platies and Guppies prefer hardish water (Guppies will, in fact, even survive in seawater if they are acclimatised properly), all three have considerable powers of adaptability and will survive and breed quite happily alongside soft-water species.

Mollies, on the other hand, have more precise water requirements, coming as they do mainly from estuarine (and therefore brackish) environments. In Florida, for instance, I have found the Sailfin, *Poecilia latipinna*, in mangrove estuaries where its co-habitants have included such distinctly salt-water organisms as Fiddler Crabs, Blue Crabs, Ribbed Mussels, Marsh Periwinkles, Eastern Oysters, Coffee Bean Shells and Soft Shell Clams.

Plate 26 Mollies, particularly Sailfins, should be regarded as brackish-water fish. (*Harry Grier/Florida Tropical Fish Farms Association*)

The water at one locality (Double Branch Creek on Hillsborough Avenue, along State Road 580) had a pH of 8, a specific gravity of 1.012 (1.020 would be true seawater) and a hardness of 200+ parts per million.

Yet, despite their requirements, Mollies are still high on the list of so-called community species, a label that implies that they will live quite happily under the same environmental conditions as Neons, Cardinals, Angels and the other tried-and-tested community species.

While *some* populations of captive-bred Molies may, in fact, do quite well in such aquaria, the majority will not. Long-term health in these fish depends on the presence of salt in the water, yet this is not suitable for the other community species and varieties.

The inevitable conclusion must therefore be that Mollies are *not* community fish, despite their worldwide popularity as such. At least, they are not community fish if the community is a traditional one, that is, if it consists of true freshwater tropical species and varieties. It can, however, be a really good community fish if the community is a brackish-water one.

Mollies, therefore, offer the aquarist the opportunity of exploring alternative tank layouts to the ones usually adopted for the other common livebearers.

The Brackish Aquarium

Brackish habitats, unlike truly marine or freshwater ones, are subjected to fluctuating salinity and temperature every single day of their existence. Not surprisingly, the only organisms found in these habitats are those that can adapt to such fluctuations – something that may come as a bit of a shock to those who believe that Mollies are delicate fish that need constant pampering to ensure their long-term survival. Sure, they are delicate – but only when we attempt to keep them under unnatural conditions.

Lying, in some ways, somewhere between marine and freshwater aquaria, brackish systems offer numerous decor possibilities. Two of the most appropriate would be the mangrove-swamp look and the marine look – the choice depending on the aquarist's preference and, ideally, on the kind of fish kept.

If the species selected for such a brackish aquarium are relatively marine in their requirements, then the best choice of layout would obviously be this one. Mudskippers, Monos, Scats, Soles and even some Sticklebacks, would be good examples of such species.

For the mangrove habitat, livebearers such as Guppies, Halfbeaks (*Dermogenys* and others), Mosquito Fish (*Gambusia affinis*), Pike-top Live-bearers (*Belonesox belizanus*) and even some Goodeids like *Ameca splendens* are suitable, although the Pike-tops and Mosquitoes are best housed on their own, because of their well-known aggressive tendencies.

Table X on page 99 lists some species of fish and plants that will do well in a brackish aquarium and will, with a few notable exceptions (all indicated), make suitable community tankmates.

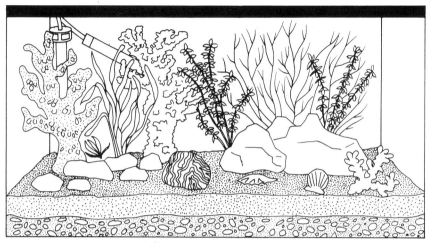

The 'marine look' brackish aquarium

The 'mangrove swamp look' brackish aquarium

Fig. 17   Two alternative brackish-aquarium layouts.

TABLE X   SOME RECOMMENDED BRACKISH-WATER SPECIES[1]

	Scientific Name	Common Names
1. FISH		
a) Livebearers	*Poecilia sphenops*	Sphenops, Green/Black Molly
	Poecilia latipinna	Sailfin Molly
	Poecilia velifera	Sailfin Molly
	Poecilia petenensis	Spiketail Molly
	Poecilia reticulata	Guppy
	Poecilia branneri	Branner's Livebearer
	Poecilia caucana	South American Molly
	Poecilia picta	Black-banded Poecilia
	Poecilia vivipara	One-spot Livebearer
	Belonesox belizanus	Pike-top Livebearer[2]
	Gambusia affinis	Mosquito Fish[2]
	Dermogenys spp	Halfbeaks
	Ameca splendens	Butterfly Goodeid
	Anableps spp	Four-eyed Fish
b) Egglayers Monodactylidae		
	Monodactylus argenteus	Mono[3]
	Monodactylus sebae	Finger Fish[3]
Scatophagidae		
	Scatophagus argus	Scat or Argus Fish[3]
	Selenotoca multifasciata	False Scat[3]
Melanotaeniidae		
	Glossolepis incisus	Red New Guinea Rainbow
	Melanotaenia spp	Australian Rainbows (several species)
Atherinidae		
	Bedotia geayi	Madagascar Rainbow
	Telmatherina ladigesi	Celebes Rainbow
Gobiidae		
	Brachygobius spp	Bumble Bee Gobies
	Tateurndina spp	Peacock Goby/Gudgeon
	Periophthalmus spp	Mudskippers[4]
	Stigmatogobius sadanundio	Knight Goby
Eleotridae		
	Dormitator maculatus	Spotted/Striped Sleeper Goby[2]
	Dormitator latifrons	Western Sleeper[2]
Cichlidae		
	Etroplus maculatus	Orange Chromide
	Etroplus suratensis	Green Chromide
	Pelvicachromis pulcher	Kribensis
	Oreochromis mossambicus	Mozambique Mouthbrooder

	Scientific Name	Common Names
Tetraodontidae		
	Tetraodon fluviatilis	Green Puffer[2]
	Tetraodon palembangensis	Figure Eight Puffer[2]
Centropomidae		
	Chanda ranga	Indian Glassfish
	Chanda woolfii	Sumatran Glassfish
Teraponidae[5]		
	Terapon jarbua[5]	Target Fish[2,3]
Toxotidae		
	Toxotes spp	Archer Fishes
Gasterosteidae		
	Gasterosteus aculeatus	Three-spined Stickleback
	Spinachia spinachia	Fifteen-spined Stickleback[3]
Soleidae		
	Achirus fasciatus	Atlantic Sole
Ariidae		
	Arius jordani	Shark Catfish[2]
Loricariidae		
	Hypostomus and *Pterygoplichthys* spp	Plecos (only some species – and at lower end of brackish spectrum)
Syngnathidae		
	Enneocampus spp	Pipefishes (some species)[3]
Lobotidae		
	Datnioides spp	Tiger Fish[2]
Cyprinodontidae		
	Floridichthys carpio	Goldspotted Killifish
	Jordanella floridae	American Flagfish
	Cyprinodon variegatus	Sheepshead Minnow
Fundulidae		
	Lucania parva	Rainbow Killifish
	Adinia xenica	Diamond Killifish
Mugilidae		
	Mugil spp	Grey Mullets (young specimens only)[3]
Percichthyidae		
	Dicentrarchus labrax	European Sea Bass (young specimens only)[3]

	Scientific Name	Common Names
2. PLANTS		
	Microsorium pteropus	Java Fern
	Sagittaria spp	Arroworts, Sagittarias
	Elodea canadensis	Canadian Pondweed
	Vallisneria spiralis	Vallis, Eel or Tape Grass
	Ceratophyllum spp	Hornwort
	Vesicularia dubyana	Java Moss
	Hygrophila polysperma	Water Star or Dwarf Hygrophila
	Echinodorus spp	Swordplants

1. Pure water has a Specific Gravity (S.G.) of 1.000. Seawater has an S.G. of 1.020 or thereabouts, the exact value depending on the particular sea or ocean from which the water comes. Brackish water falls somewhere in between, but it is impossible to give it an exact, single, inflexible value in terms of S.G. Any water having an S.G. value of between 1.005 and 1.015 can be regarded as brackish. An average value of around 1.008–1.010 would provide adequate living conditions for most of the species listed in the Table. Brackish conditions can be achieved through the addition of as little as 5 ml (1 teaspoonful) of salt (preferably a marine mix) per 4.5 litres (1 Imperial gallon) of aquarium water.
2. Species in this category are undesirable for mixed communities because of their aggressive or predatory tendencies.
3. Species in this category are better suited to brackish environments near the marine end of the scale.
4. Mudskippers need an area of moist land on to which they can climb and on which they can feed.
5. The usual spelling found in the aquarium literature for this family is Theraponidae, with this species appearing as *Therapon jarbua*. However, the original spelling on record for the genus is *Terapon*. Therefore, although *Terapon* represents an incorrect transliteration of the Greek word (for 'slaves') from which the name was derived, the strict rules of nomenclature dictate that the original spelling should stand as the valid one, hence Teraponidae and *Terapon*.

Aquarium Layout

The eventual aquarium layout one chooses depends to a very large extent on the purpose for which the particular aquarium is being set up.

The layouts suggested here, for instance, have a particular purpose – they are intended for brackish water *communities* of fish.

If individual species of livebearers are being kept, then this too will have an influence on the overall design. The choice of layout will then depend on whether the aquarium is being set up purely as a decorative feature in which appreciation of the fish is paramount, or as a working/breeding set-up where the main purpose is to encourage reproduction.

Display Tanks

In set-ups designed for display, decorations and plants arranged around the periphery, with a clear open area in the centre foreground, will show off the fish beautifully. This principle can, of course, be applied equally both to brackish and freshwater systems.

A shoal of fancy, colourful Guppies, for example, can look quite stunning in such an aquarium, particularly if the plants chosen complement the fish. For the Guppy aquarium, the plant most often recommended is the Indian Fern (*Ceratopteris*), also known as the Water Sprite. Its bright green, wide fronds seem to show off the spectacular colours of a shoal of male Guppies almost to perfection, particularly in a well-lit aquarium.

Plate 27   A well-designed, planted and stocked community aquarium forms a spectacular focal point in any living room. (*Bill Tomey*)

One advantage of using this plant is that it does well in water conditions that suit Guppies (alkaline and slightly to moderately hard). Of course, the compatibility of the chosen plants with the water conditions required by the fish species is another factor that should play a part in determining the aquarium layout. Swordtails and Platies also enjoy similar conditions, although Indian Fern may be a little too tender to withstand the somewhat more energetic grazing behaviour of these species in the long term. Mollies, if they are to be kept in freshwater aquaria, also like hard alkaline water, although they prefer slightly higher temperatures.

In general, the sort of layout discussed above will prove suitable for most species of livebearer, with minor modifications to take account of specific requirements – the Pike-top Livebearer (*Belonesox belizanus*) for example, would appreciate dense cover in the corners of the aquarium.

The size of the fish should always be taken into consideration when selecting an aquarium. Small species, such as *Heterandria formosa*, one of the Mosquito Fishes, and one of the smallest vertebrates known to science, tend to get lost in large, densely planted aquaria. Yet, given moderately sized accommodation with adequate clear viewing areas, and moderate lighting, the beauty of these fish will be fully appreciated. The relatively large size of the new-born fry, added to the predominantly non-cannibalistic nature of *H. formosa* and other small peaceful species, also means that an aquarium set up primarily to show off the fish will serve equally well for breeding purposes.

Some small types of livebearer – such as *H. formosa*, or *Quintana atrizona* (the Black-barred Livebearer) – prefer quiet-water conditions, while others – such as *Phalloptychus januarius* (the Barred or Striped Millions Fish) – prefer highly aerated water with some movement, although they are also found in polluted, oxygen-depleted waters in Rio de Janeiro. Aquaria set up for these fish should obviously take their differing requirements into consideration.

While small aquaria – say, 45 × 25 × 25 cm (18 × 10 × 10 in) – for small species may look very attractive, they do require more careful management than larger set-ups. The main source of the majority of problems is the small volume of water involved. The larger the volume of water, the greater its overall stability, and, consequently, the greater its capacity to cushion or buffer the adverse effects of an oversight or mismanagement on the part of the aquarist. Large volumes lose heat, become polluted, and fluctuate in quality more slowly, and to a lesser extent, than small volumes. Therefore, while a large aquarium may be able to assimilate the potentially adverse effects of a single, accidental case of overfeeding, a small one cannot.

Among the fish that require special tank layouts for their qualities to be fully appreciated are the Four-eyed Fishes (*Anableps* spp). These brackish-water fish spend most of their time swimming just under the surface, with the top half of their eyes exposed to the air, frequently ducking their heads to keep their eyes moist. They also come out on land, almost like Mudskippers, either simply to lie on moist sand or mud, or to feed. Provision should therefore be made to allow them to exhibit this interesting aspect of their

Plate 28   Even if an aquarium is being set up with a single species in mind (such as these *Limia nigrofasciata*) due attention must be given to layout and design. (*Bill Tomey*)

behaviour. At least one public aquarium feeds its *Anableps* by placing their food on a square sheet suspended just above the surface of the water, thus forcing the fish to come on to dry (or, at least, moist) land to feed.

The trouble with creating a gentle sand or gravel slope or beach in an aquarium is that both gravity and the disturbance generated by the fishes' activity will soon level out the sand or gravel. For obvious reasons, mud cannot be used, either. The best approach is to build a series of terraces, the topmost of which reaches to just below the water surface. The top of this terrace can then be sloped in beach-like fashion and will only require minimal maintenance to retain its shape.

The construction of a beach also dictates that the water level in an *Anableps* tank be kept lower than in more traditional livebearer aquaria. Lowering the level has the added advantage that it allows the aquarist to appreciate the four-eyed characteristic of these fish to the full.

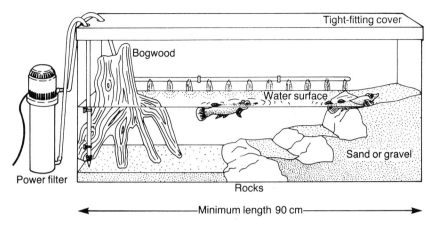

Fig. 18   An aquarium for Four-eyed Fish must take their very special requirements into consideration.

A good aquarium cover performs two useful roles. It prevents the fish from jumping out (Four-eyed Fish are excellent jumpers) and keeps the air above the water humid just as it is in the wild.

Four-eyed Fishes are large compared to most other livebearers. *Anableps anableps*, for instance, can grow to 30 cm (12 in). This, added to a predominantly carnivorous diet, consisting largely of insects in the wild, means that these fish are very efficient water pollutors. Power filtration is, consequently, a must.

Shallow, brackish water virtually excludes live plants, at least on a long-term basis. However, this does not necessarily mean that an *Anableps* aquarium need be devoid of decorations. Driftwood (properly treated with a polyurethane varnish to prevent rotting or excessive leaching out of tannins), especially if arranged so that it projects above the water surface to simulate mangrove-tree roots, will provide an apt and attractive setting.

An *Anableps* species tank should be *long*, to provide a shoal with a sufficiently large area of water surface in which they can swim in formation. An aquarium 90 cm (36 in) long should be regarded as the absolute *minimum* size for these species. Most standard aquaria of this length are 38 cm (15 in) deep and 30 cm (12 in) from front to back and, since depth is wasted on *Anableps*, it would seem sensible to construct an aquarium specifically for them. I would suggest that such an aquarium should be at least 120 cm (48 in) long, 30 or 38 cm (12 or 15 in) deep, and 38–60 cm (15–24 in) wide, with a purpose-built cover to fit, possibly consisting of two halves that overlap and can be slid open for feeding and maintenance.

An aquarium set up like this, primarily with display purposes in mind, will often prove adequate for breeding purposes as well. Experienced or specialist livebearer keepers will, however, often adopt quite a different approach with their working or breeding tanks.

Working Tanks

A working tank may be loosely described as one whose primary aim has more to do with the keeping of fish for their own sake (or that of the fishkeeper), for example for study, rather than for display. One is therefore much more likely to find working tanks in specially constructed, or adapted, fish rooms or fish houses. Working tanks are also much more likely to be owned by established enthusiasts rather than fledgling aquarists.

These specialists will tend to own more aquaria than beginners, sometimes as many as a hundred or more. It would obviously be extremely expensive to set up such a large number of aquaria if they were all fully equipped with attractive hoods, if each had its own decorative stand or cabinet, and if each contained a full complement of plants, gravel, etc. Maintaining them in good decorative order would also prove extremely time-consuming.

Working tanks are, therefore, considerably more basic than this. Yet, they do not overlook the biological needs of the fish. Many specialist aquarists will use gravel in their working tanks, but an equal number will dispense with gravel altogether. While the latter approach has little to recommend it in terms of aesthetic value, it does make water changes and overall cleanliness a lot easier.

Shelter for the fish may be provided by natural or artificial bogwood or caves, plastic plants, clumps of Java Moss (*Vesicularia dubyana*) or any other appropriate means.

Aquarists who take their interest in livebearers to these levels will not (*cannot*, in fact) be satisfied with a single attractive aquarium which forms part of their furniture, contains a mixed community of fish, and is admired from a distance. They may, indeed, well have such an aquarium, but they also require more from their hobby – much more.

They want to know as much as possible about their fish, from the highly individual and often minute physical and behavioural differences that exist between closely related species, or geographically isolated populations of the same species, to their optimal temperature, chemical and nutritional requirements, their susceptibility to diseases, stress, cannibalism or whatever. In fact, these dedicated aquarists have an almost insatiable desire to learn about their fish, and livebearers are perfect subjects for them.

Their (generally) small size means that even a modest-sized room can accommodate large numbers of aquaria, especially if they are arranged in racks and are placed end-on. Most, if not all, of these aquaria will be species tanks – they will contain a single species of fish, often only a single pair.

Although each aquarium may contain a heater-thermostat, it will be there only as a safety precaution (or for fine environmental control) because fish rooms or houses are normally space heated. In other words, the whole room is heated, often via an extension of the house's central-heating system.

Centralised air supplies are also popular among specialists, their designs usually being similar to those found in most aquarium shops. Filtration, though, tends to be arranged on an individual-tank basis, which allows for more precise control than that afforded by a centralised system.

Illumination is often provided by means of household fluorescent tubes attached to the underside of each shelf rack and working on automatic timers, providing an approximate daily cycle of 12 hours on and 12 hours off, thus reflecting the day/night periodicity found in the tropics and subtropics where most of the livebearers originate.

Working tanks differ from display tanks in many other ways, some as easily discernible as those outlined above, some less so. In the latter category, I would place aquaria that are sometimes used as nursery tanks and which contain a mixture of compatible, similarly-sized, non-hybridisable juvenile livebearers. These aquaria, often containing a full complement of gravel, plants and natural decorations, are attractive as well as functional.

Breeding Tanks

It is quite difficult to draw a clear-cut dividing line between a working/display tank and a breeding tank. The main reason for this is that some working/display layouts can work just as well for breeding purposes for species that are not particularly cannibalistic. Some such species have already been mentioned, but there are numerous others, particularly among the Mollies, Limias and their immediate relatives. Even some predatory-looking species like the Butterfly Goodeid (*Ameca splendens*) can be kept in mixed adult/fry associations for most of the time.

Despite this, it is generally better for a gravid female to drop her fry in peace rather than in the hustle and bustle of a tank holding other fish. This is mainly because of the males of the species, most of which exhibit heightened interest in a female as soon as she enters the later, final, pre-birth stages and, especially, once birth is underway.

The most commonly encountered problem with isolating a female in a tank on her own comes from the difficulty of estimating the correct amount of food that needs to be provided. A second problem sometimes encountered arises from the tendency that some females (particularly those belonging to shoaling species) exhibit towards sulking in the farthest corner of the tank.

The former problem is often overcome by experienced fishkeepers simply through the introduction of aquatic livefoods (usually *Daphnia*), which will survive without the risk of water pollution.

The second is often tackled through the provision of company, in the form of a similarly gravid female of the same species or of a small shoal of juvenile fish that are too large to be eaten by the female and too small to prey on the newly born fry when they arrive.

In either case, several aspects of tank layout will help further. If the tank is bare – if, that is, it has no gravel – then the introduction of a breeding mop (as used in breeding Killifish), either commercially produced or home-made from strands of undyed wool several inches long, tied to a piece of cork, can provide shelter, not just for the female but for her fry as well. Undyed nylon pot scourers, rinsed and partly teased out, will do a similar job, as will a few clumps of Java Moss (*Vesicularia dubyana*), bunches of coconut fibre, clumps of prepared Spanish Moss (a Bromelliad, or air plant, found in great abundance in the southern United States), or any other non-toxic fibrous equivalent. In addition, there should always be a generous surface layer of a floating plant such as *Riccia fluitans* (Crystal-wort). Failing this, the fine feathery roots of larger plants like the Water Hyacinth (*Eichhornia crassipes*) or Water Lettuce (*Pistia stratiotes*), should be provided.

If the aquarium has gravel in it, a well-established carpet of that delight-ful dwarf plant *Lilaeopsis novae-zealandiae* will provide excellent shelter at ground level for the fry of those species that are not surface swimmers.

Whether the tank is bare or not, a breeding set-up should incorporate power filtration only if it is capable of being turned down to a gentle flow. Particularly popular among livebearer specialists are internal air-operated box and foam/sponge filters. The air lift is usually curved to allow for the exhalent stream of water and bubbles to be directed along the surface of the water. This characteristic can be beautifully exploited in livebearer breeding aquaria to create a water current that will gently direct newly born fry towards a pre-selected area of safety.

By far the best of such systems I have encountered was devised several years ago by one of the top livebearer hobbyists in the United Kingdom, Dennis Barrett, a founder-member of the Southern Livebearers Aquatic Group, one of two specialist societies in the country (the other one being the excellent Viviparous – the Livebearer Information Service). The approach is simplicity itself. Two sheets of glass are stuck vertically inside a tank of any dimension, but usually around 45 × 25 × 25 cm (18 × 10 × 10 cm), to form a broad V-shape with a small gap at the apex of the V. This gap is large enough to allow fry, but not adult fish, through. The larger part of the tank – that is, the part 'above' the V – is set up with gravel, a few (not too many) plants, a heater-thermostat, and a sponge/foam air-operated filter. The curved outlet from this filter is located more or less flush with the surface of the water, with the current being aimed at the apex of the V. The portion of the tank below the apex of the V is left bare. The gravid female is then placed in the furnished (larger) part of the breeding tank with an ample supply of livefood.

As the fry are born, the gentle current, plus the fry's own instincts to swim away from predators (in this case, their own mother) will result in a relatively high number of them ending up within the bare, and safe, section of the breeding tank where they can be left, and fed, until they are large enough to be moved to alternative quarters. By keeping this section of the tank bare, any wastes produced as a result of feeding, either with livefoods like newly hatched Brine Shrimp (*Artemia*), or other fry foods, can be quickly and effectively removed, thus minimising the risk of water pollution.

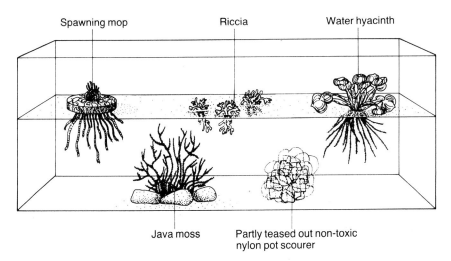

Spawning mop   Riccia   Water hyacinth

Java moss   Partly teased out non-toxic
nylon pot scourer

Fig. 19 Breeding tanks can vary enormously. Two possibilities are illustrated – one for a bare set-up, the other (below) for a furnished one.

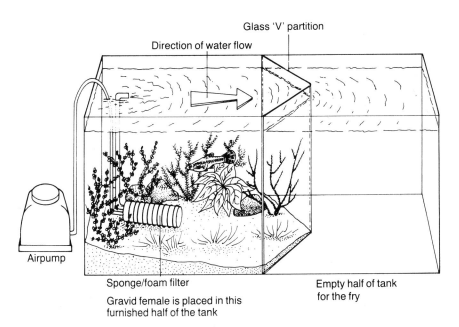

Glass 'V' partition

Direction of water flow

Airpump

Sponge/foam filter

Gravid female is placed in this
furnished half of the tank

Empty half of tank
for the fry

Fig. 20 This ingenious and highly effective breeding tank was designed by Dennis Barrett, a leading UK livebearer specialist.

Breeding Traps

The term breeding trap is usually reserved. for commercially produced plastic mini-tanks or containers in which gravid livebearer females can be placed when they are about to give birth.

The trap is generally divided into two unequal halves. The top, and usually larger, half is designed to hold the female, while the lower half will eventually hold the fry. In most models, the two halves are separated from each other by a shallow V-shaped partition in which the apex is directed downwards. A narrow slit separates the two arms of the V, in much the same way as that found in the breeding tank described in the previous section, but oriented horizontally.

Once parturition has been completed, the female can be removed to be returned to her home tank or, preferably, to a separate aquarium to give her a chance to recuperate.

The fry, meanwhile, are left in the trap. By removing the partition – a facility that is incorporated into many designs – the fry are, effectively, provided with a small, safe aquarium until they are large enough to be released into the main tank, or until alternative, larger quarters can be provided.

Breeding traps have small, external flotation chambers to allow them to float inside a larger aquarium. They therefore require no additional heating.

Plate 29 There are numerous breeding trap designs available. These are particularly useful for single-tank owners. Also shown, a hatcher for Brine Shrimp – an ideal early food for livebearer fry. (*Interpet Ltd.*)

As a rule, breeding traps are left bare, although some aquarists will add a little functional decoration in the form of a few floating plants. Their small size does not really allow for any form of filtration equipment to be installed, though, and this is perhaps one of their major drawbacks.

As discussed earlier, small volumes of water are highly susceptible to quality fluctuations. Breeding traps, being usually smaller than even the smallest aquaria, are particularly prone to this potentially dangerous problem. They therefore need careful handling, especially when it comes to feeding. Aquatic livefoods, adequately sifted to remove potential fry predators, such as small damselfly nymphs, can be substituted for dry, freeze-dried or deep-frozen foods, if the aquarist is unsure of the correct amount to provide for a single female or a batch of tiny new-born fry.

In addition, solid wastes and small partial water changes should be carried out on a very regular basis (perhaps, as frequently as once a day) even in those breeding traps whose floor is perforated to allow some degree of interflow between the trap water and that in the aquarium.

Despite these drawbacks, breeding traps provide the single-aquarium owner with the opportunity of breeding fish and rearing at least some of the offspring – something that would otherwise be next to impossible.

Established or multiple-tank hobbyists generally adopt a range of different approaches, some of which have already been described in earlier sections.

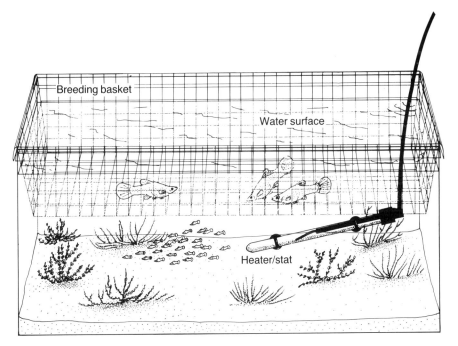

Fig. 21 The breeding basket provides all the advantages of a breeding trap, plus a few extra ones.

One additional technique is worth mentioning here because it represents a modification of the breeding-trap principle that has been found to be particularly effective. The breeding trap in this case is a shallow flat-bottomed basket made of plastic-coated medium-sized mesh netting (the size of the mesh being large enough to allow the fry through but not an adult fish). The size of the basket is a matter of personal choice but, generally speaking, it is slightly smaller than the breeding tank itself. The floor of the tank can be either bare or furnished with gravel and some short plants such as *Lilaeopsis novae-zealandiae*. Once the tank has been filled with mature water (or water that is allowed to mature), and a heater-thermostat and filter are installed, the basket is suspended from the edge of the tank (with string, non-toxic plastic-coated wire, or a suitable equiva-lent) so that its top edge projects above the water surface by 2.5 cm (1 in) or, preferably, more. The gravid female is then released into the trap and fed normally. When she eventually drops her fry, these will swim through the mesh to safety. The female and basket are then removed and the fry allowed to grow on in the tank.

There are several significant advantages with this method, which has been found particularly effective when several females, either of the same or of compatible (but non-hybridisable) species are due to give birth at around the same time. For a start, the fact that a whole tank is being made available (the mesh will not impede water interchange) minimises the risk of water pollution. Then, the exceptional effectiveness of the mesh means

Plate 30 In some Far-Eastern farms, the breeding pond is used extensively in livebearer production. A vertical mesh separates parents from their offspring. (*John Dawes*)

that the fry have no difficulty in finding a route into the main tank. Consequently, large numbers can be saved in this way. Further, the sulking that sometimes occurs when a single female is removed from a populated tank and kept on her own until she gives birth can be easily avoided by keeping several females, gravid or not, together inside the trap.

Once the adults and basket are removed, we end up with an already established, furnished tank which only needs (perhaps) the addition of a few extra plants. The fry, consequently, do not have to undergo the stresses associated with being transferred to other quarters and, therefore, end up with one of the best starts possible.

This technique has been found to be so successful that many of the professional breeders who produce their stocks in (bare) aquaria regularly employ this type of trap. Even those commercial enterprises that produce their fish in outdoor earth ponds (as they do in the tropics) use this technique, albeit with giant baskets, more appropriately called cages, suspended from stakes, and mixed male/female broodstock.

Where concrete vats/ponds/troughs are used, the trap principle is sometimes adapted into a vertical partition that separates the mixed male/female broodstock from their offspring.

Compatible Tankmates

Most livebearers are, generally speaking, quite adaptable creatures. In fact, many have such great adaptive powers that they will survive conditions that would kill other fish. Guppies and Mosquito Fish (*Gambusia affinis*) are perhaps the toughest of the lot, but there are quite a few others as well.

This tough reputation has, unfortunately, led to many livebearers being kept under less than favourable conditions – forced to occupy aquaria containing totally unsuitable tankmates and having both unsuitable layouts and unsuitable water chemistry.

Tiger Barbs (*Barbus tetrazona*) and Angels (*Pterophyllum scalare*) have already been singled out as attractive fish which can nevertheless present dangers for some livebearers. This aside, most of the commonly available livebearers and community tropicals can co-exist quite happily with a wide range of other, lesser-known species.

Even for livebearers that should be kept away from other surface or midwater swimmers, some compatible tankmates can be found, mainly among the large range of docile bottom-dwelling fish such as *Corydoras* and *Brochis* catfishes, or the various Kuhli Loaches (*Acanthophthalmus* spp).

If the main purpose for keeping livebearers is display, rather than breeding success, then the range of compatible species that can be kept is enormous. Any species of fish that will do well in the same water conditions may be kept, just as long as behavioural characteristics do not dictate otherwise.

If breeding success is a priority, however, then the situation is quite different. Not only must potential fry predators be kept out, but also any species and varieties of livebearer that are likely to affect the outcome of any breeding attempt.

Top of the list of exclusions are all species and varieties related to those one hopes to breed. The main reason for this exclusion is the propensity that livebearers exhibit for hybridisation, particularly in the confines of an aquarium, where behavioural barriers that would normally isolate the fish in nature become eminently surmountable.

There is, of course, an added problem when it comes to livebearers (as opposed to egglayers). Females of many types of livebearing fishes, particularly those of Poeciliid livebearers, are capable of storing sperm for considerable periods. As a result, a single insemination will lead to a series of broods spaced out over a number of months. If the sperm comes from the appropriate male, then this is fine. However, if the inseminating male comes from a related species or variety, then the female can be considered worthless as far as maintaining the purity of the species or variety is concerned. Obviously, though, as a living, feeling organism she is just as valuable as any other fish and should therefore receive equal care and attention, even though she must not be used as part of any future breeding programme.

The ease with which some species of livebearer will hybridise is quite exceptional. The best-known hybridisations are those between Swordtails and Platies (*Xiphophorus helleri* and, usually, *X. maculatus*). These fish are so closely related genetically that their hybrids are fully fertile, a factor that has been exploited to such an extent over the years that few, if any, pure Swordtails and Platies are now available to the general hobbyist.

This is perfectly acceptable from the non-specialist point of view, but obviously not so far as specialist hobbyists are concerned. They require not just pure species, but also pure *strains* from different localities within their natural ranges. Clearly, when one gets down to this level of specialisation, the degree of caution that must be exercised is extreme – and it is thanks to the dedicated efforts of these aquarists that there are currently over 150 pure types of wild livebearers being kept and bred in captivity.

The significance of this fact cannot be over-stressed, especially in the cases of those species or varieties that are under threat in the wild. In fact, in at least one case, the Sawfin Goodeid (*Skiffia francesae*), it is only through the concerted efforts of aquarists on an international level that the species has survived, all known natural populations having been exterminated through the introduction of Platies into the species' only recorded habitat.

As Langhammer has demonstrated, *Skiffia francesae* is quite capable of hybridising with at least one of its close relatives, so due care must be exercised by all who are fortunate enough to have the species in their aquaria.

One saving grace that Goodeids possess over Poeciliids, should an unforeseen hybridisation occur, is that Goodeid females cannot store sperm. Therefore, once the hybrid brood has been produced, the female can be re-incorporated into future breeding programmes.

To return to the Poeciliids, even species that are more distantly related to each other in genetic terms than are Swordtails and Platies will hybridise. Guppy × Molly hybrids, for example, are not uncommon, although they all seem to suffer viability problems of one kind or another – an indication that their genetic make-ups, while being sufficiently similar to each other to permit fully formed offspring to develop, are nevertheless different enough for some degree of incompatibility to become apparent.

Many Limias, like Swordtails and Platies, will not only hybridise but will produce fertile hybrids. Some experiments I carried out with *Limia melanogaster* (the Blue Limia) and *L. vittata* (the Cuban Limia) produced very interesting results indeed. The hybrids in question, although appearing roughly halfway between both parental species overall, tended to look more like the mother than the father. So, if the female was a *L. vittata*, the hybrids seemed to exhibit more spots than those where the female was a *L. melanogaster*, these latter hybrids being also somewhat bluer. This is probably one of those relatively rare examples of *cytoplasmic inheritance* in which factors other than those contained exclusively in the genes make some contribution towards the make-up of the organisms concerned. Since, at fertilisation, females contribute yolk in addition to genes, while the males' contribution is virtually exclusively genetic, any extra-genetic influences exhibited by the offspring tend to be maternal in nature. Perhaps some researchers will be able to investigate this occurrence in *Limia* in greater detail someday. Whatever the parentage of these *vittata* × *melanogaster* hybrids, they proved fertile up to the third generation, during which a major electrical power failure caused the death of all my stocks and put paid to further research.

Another experiment I carried out highlighted the potential dangers of hybridisation even further. I had been keeping a Golden/Red-eyed female Sailfin Molly (*Poecilia velifera*), which I presumed to be virgin, with a Black Sailfin male (probably not a pure *P. velifera*). Eventually, the female dropped three fry which, in time, developed into Green Sailfin females – perhaps indicating that she was not a virgin, as I assumed, but had been mated by, probably, a Green Sailfin male at some stage prior to the attempted Black/Gold cross. I subsequently placed the three young (definitely virgin) female Green Sailfin Mollies in an aquarium with a single hybrid *Limia melanogaster* × *nigrofasciata* male, the sole survivor of an earlier hybridisation attempt. Some eight weeks later, one of the females dropped two fry. Only one survived to adulthood, ending up as a somewhat Molly-like male which was unable to maintain an even keel and always swam with his head pointing upwards at approximately 45°.

One of the several significant things that the above shows is that, as long as a hybrid male is fertile, he may be capable (under certain circumstances) of inseminating a female of a species other than the two from which he originated – possibly even when the inseminated female is, herself, of mixed origin.

Despite this quite astonishing fecundity, the fact that small numbers of fry were produced and that, in the last instance, the resultant hybrid male was not fully viable (in fact, he failed to inseminate any of the virgin Molly females that I introduced into his aquarium over a period of three months), shows that many hybridisations will, sooner or later, lead to a dead end in reproductive/evolutionary terms.

This is not, however, always the case. With Swordtails and Platies and, at least, with two of the *Limia* species I experimented with, the fry are often larger, more colourful and reproductively more vigorous than their parental species. This hybrid vigour, or *heterosis*, has, not surprisingly, been exploited commercially with most of the popular species and with varieties of Poeciliid livebearers to produce ever more colourful and often larger fish.

Extra-vigorous hybrids can out-perform their pure counterparts. For instance, in 20 experimental attempts, a single hybrid *Limia melanogaster* × *vittata* male invariably managed to mate with a *L. melanogaster* female sooner than either *L. melanogaster* or *L. vittata* males introduced into the experimental tank at the same time (the quickest insemination took just 15 seconds!). Clearly, super-vigorous hybrids like this can wreak havoc in a population of otherwise pure fish should they, intentionally or accidentally, be allowed access to receptive females.

Specialist livebearer hobbyists and their associations or societies therefore tend to take extra-meticulous steps to ensure that hybridisations are

Plate 31   Hybrids can often prove more vigorous than their parental species and can wreak havoc among pure stocks. The male seen mating with a Blue Limia (*Limia melanogaster*) female is a *L. melanogaster* × *vittata* hybrid which was invariably quicker at securing a mate than males of either of its parental species. (*John Dawes*)

avoided. These steps may include collation of details of which members own which species and/or varieties, the exact origin of these stocks, detailed bloodline family histories of the stocks, records of transfers between members, and so on. As a result, several species that are currently known to be under threat in the wild have a fairly assured future under the responsible ownership of these aquarists, who, through concerted and controlled breeding programmes, are playing a vital role in wildlife conservation.

To sum up, it is probably reasonable to regard most livebearers as being *incompatible* with each other, except where known facts indicate otherwise – for example, as far as I am aware, no hybridisations have ever been recorded between any of the *Xiphophorus* species and their *Poecilia/Limia* counterparts.

I would suggest that the best way to avoid all risk of accidental hybridisation is to regard all livebearers as potentially guilty unless proved innocent in the first instance, changing this verdict only on the advice of experienced specialists who will be able to say which fish can be safely housed together and which cannot.

Nutrition

With a little commonsense, most livebearers can easily be provided with a suitable diet in captivity. Obviously, individual requirements and habits, such as the predatory lifestyles of *Belonesox*, Halfbeaks and some *Gambusia* species, need to be taken into consideration, but, generally speaking, livebearers are omnivores.

This means that they will normally take both animal- and plant-based foods. This applies even to those species that are regarded as herbivores (plant eaters), like many of the Mollies. In fact, Mollies will sometimes eat their own fry, proving beyond doubt that they can digest animal protein just like any carnivore.

Having said this, though, many livebearers enjoy a diet that is biased towards the vegetarian end of the nutritional spectrum, so this should be borne in mind when deciding the menu. (Some guidance about this is provided in the species entries in Part III.) The overall menu should, of course, consist of a balanced diet – one that will provide proteins, carbohydrates, lipids (fats/oils), vitamins, mineral salts (trace elements), and a bit of roughage.

These compounds perform the same roles in livebearers as in other fish, ranging from the growth and repair of tissues to the maintenance of life itself. Therefore, during periods of growth or recuperation, diets that are slightly higher in proteins than normal (say, around 35%), may be found advantageous. At other times, lower levels of protein (around 30% in most commercial flake foods) will suffice.

Even this lower figure will provide livebearers with more protein than they require, but is, nevertheless, a satisfactory way of ensuring that no protein deficiency will arise out of the use of commercially prepared diets.

Fish-food manufacturers claim that their products will meet all the dietary requirements of fish, even for breeding purposes, and the evidence tends to support this. Even so, the majority of experienced livebearer keepers provide their fish with livefoods in addition to, or in some cases instead of, dry prepared foods.

There is no doubt that the vast majority of livebearers seem to enjoy chasing after livefood, and this may indeed be so. However, unless due care is taken, parasites, other pathogens and predators can be introduced into an otherwise healthy tank, particularly if the food has been collected from the wild. Despite the possibility of this happening, I tend to feel that the dangers of using collected livefoods are somewhat over-stressed in aquarium literature. Certainly, it is possible to introduce disease organisms with aquatic livefoods, and certainly – at least where endangered species are concerned – the risk may not be worth taking. However, healthy fish have considerable in-built resistance to pathogens, so it may well be that livefoods (even wild-caught) may not be as dangerous as they are sometimes made out to be. Nevertheless, if one wants to make sure that no 'nasties' inadvertently find their way into the aquarium, then the answer is to restrict the choice of food to commercially prepared formulations, clinically cultured aquatic livefoods or non-aquatic livefoods such as fruitflies and earthworms.

If the commercial option is taken up, then the choice available nowadays is quite staggering – and becoming more so with each passing year. Flake, powdered, granular, pelleted, 'stick', freeze-dried, deep-frozen/irradiated, micro-encapsulated and several other types, will ensure that virtually every need can be met.

Among the dry (flake/powdered/granular/pelleted/'stick') categories, variations in composition, from predominantly vegetable-based recipes to predominantly flesh-based ones, allow a degree of flexibility and precision of choice that was unavailable to most aquarists as recently as fifteen years ago.

On the livefood front, at least the following are regularly or occasionally used by aquarists: Infusoria (usually *Paramecium*), Green Water (mainly *Euglena*), Rotifers, newly-hatched Brine Shrimp (*Artemia*), adult Brine Shrimp, *Mysis* shrimps, freshwater shrimps (e.g. *Gammarus*), Fairy Shrimps (*Streptocephalus seali*), *Daphnia*, *Moina* (a small water flea used in the tropics), *Cyclops*, *Tubifex* and Black worms (a mixture of Tubificid species), Whiteworms (*Enchytraeus albidus*), Grindal Worms (*Enchytraeus buchholzi*), Microworms/eels (*Anguillula [Panagrillus] silusiae*), Vinegar Eels (*Turbatrix aceti*), Mosquito and Midge larvae and pupae (several species), Glassworms (*Chaoborus*), Bloodworm (Chironomid larvae), Fruitflies (*Drosophila melanogaster*), Mealworms (mainly *Tenebrio*, *Blapstinus* and *Alphitobius*), chopped/mashed Earthworms (usually *Lumbricus terrestris*) and *Allolobophora longa*.

Details of how to culture these and other livefoods are available but are, obviously, outside the scope of this book. For further details, the reader is guided to one or other of the articles and books listed at the end of this chapter.

Faced with problems of seasonality, of local availability, or of difficulty of culture, the majority of hobbyists turn from livefoods either to commercially prepared foods or to home-made pastes made of beefheart, lettuce/spinach/peas, and other combinations.

In the end, what everyone needs to aim at is providing a diet that meets all the nutritional requirements of the fish. For this reason, it is always advisable to vary the menu on a regular basis, thus ensuring that what one particular food item may lack will be supplied by another. Where top-quality commercial foods are concerned, they have been specifically developed to provide all the essential ingredients, but even so variety does appear to have a beneficial effect on the fish, which tend to respond with relish to dietary changes, particularly after a lengthy run on the same type of food.

Assuming that all other parameters are in order, a well-fed fish is usually a healthy fish. Should one or other essential ingredient be overlooked, however, then problems will undoubtedly arise sooner or later.

Perhaps the only effect will be a reduction in the rate of growth, or perhaps a slight reduction in the number of fry produced in a single brood, or in their size at birth, or some other deviation from the norm so small that only the experienced aquarist will notice that anything is amiss. Conversely, the effect could be so marked – for example, the development of a bent backbone (Scoliosis) – that no-one could fail to notice the deficiency (in this case, of Vitamin C – ascorbic acid – which is associated with collagen synthesis and is also found in bone and scales).

Metabolic processes, from the formation of visual pigments, to the oxidation of sugars, proper functioning of the nervous system, energy release, fat/carbohydrate/protein metabolism, production of pigments, formation of blood-clotting factors, respiratory reactions, absorption of calcium, phosphorus and other minerals, plus numerous other reactions, would be impossible in the absence of the appropriate vitamins.

Therefore, even if there was enough calcium around in the diet to build a caveful of stalactites, it would be totally useless to a fish if the necessary vitamins (such as Vitamin D which aids the absorption of calcium from the intestine) are absent.

Generally speaking, fish are overfed, rather than underfed. For this reason, they can be left unfed for some time without any adverse effects. In fact, some aquarists adopt a six-day feeding/one-day fasting weekly programme. During the day of fasting, the fish will use up some of the surplus fats, oils and carbohydrates they may have accumulated during the other six days, but if they wish to feed, then a well-established, planted aquarium will provide them with snacks in the form of algae, small snails, crustacea, micro-organisms and other titbits.

Indeed, as long as a well-balanced feeding regime has been adhered to for some time, fish can be left for several weeks without additional food, during which time they will use up some of their reserves, feed on the bits and pieces mentioned above and, in a well-balanced planted set up, generally manage very well.

The fact that many livebearers are omnivorous, and will therefore eat algae and bits of vegetation, makes them ideal pets for the aquarist who is regularly involved in travelling. Even so, if it is felt desirable to provide food during such absences, then commercially available automatic feeders will be found quite useful. Weekend or vacation blocks, in which food fragments are embedded in a slow-dissolving matrix are also available. Should these be used, though, a partial water change is advisable on returning home in order to redress the rise in pH and hardness that results from the use of these blocks.

Any of the above suggestions is preferable to leaving the fish in the care of a neighbour or friend, however well-intentioned he or she may be – that is, unless the person in question is an aquarist who understands the ins and outs of feeding fish, or is someone you have trained in the right techniques, or unless you leave pre-prepared measured amounts of food for each aquarium for the time you are likely to be away.

Health Care

As long as aquarium conditions are good and fish are provided with a nutritious, well-balanced diet, they will tend to remain healthy. There is, in fact, no doubt that most disease problems can be avoided simply through a commonsense approach to aquarium management.

Conversely, poor conditions, inappropriate diet or mismanagement will cause stress, with all its associated problems, perhaps the most damaging being that the fish's resistance to pathogenic (disease-causing) agents can be lowered to such an extent that they succumb to some ailment or other.

Livebearers, of course, are not exceptional in this respect, even though many species are quite tough and can withstand conditions that other, more delicate, types would find intolerable. Prevention is, without doubt, immeasurably better than cure – but it isn't 100% foolproof. Therefore, sooner or later, health problems will arise.

When they do, livebearers are, in general terms, no different to other fish and can suffer from the same wide range of diseases. In addition, though, there are several problems that seem to be associated with live-bearers more frequently, or affect them more seriously, than other fish.

Fin Rot

Fin Rot is a disease that, as its name implies, causes erosion of fin tissues. In the vast majority of fish, treatment will result in the fins re-growing, not always to their full, former glory, but usually pretty close to it.

In Poeciliid, Anablepid and Hemirhamphid livebearer males, however, the anal-fin modifications make complete regeneration somewhat more difficult. The result can be that males affected by Fin Rot may not be able to mate successfully, at least, until their gonopodia (in Poeciliids and Anablepids) or andropodia (in Hemirhamphids) have re-grown fully. Even then, it is possible that their effectiveness will be permanently impaired.

Belly Sliders

When fry are born, one of their first 'biological duties' is to surface and fill their swimbladders with air. Once this is done, they can control their buoyancy and swim effortlessly at any level of the water column. Live-bearer fry appear to suffer from an inability to inflate their swimbladder somewhat more frequently than egglayer fry – or it may be that, because of their larger size and lower numbers, affected fry are more obvious.

Belly sliders appear to be more common among the fry of females producing their first broods, particularly if the females themselves are on the young side. Broods trapped in deep water, for example, in heavily planted aquaria where the fry find it difficult to surface, or in particularly deep aquaria, also seem to be susceptible to this problem.

Transferring the affected fry to a shallow tank and raising the temperature may help, but, generally speaking, the belly-sliding condition proves to be unrectifiable. Humane disposal – preferably using a suitable anaesthetic like MS222, obtainable from a vet – would seem to be the kindest action to take.

Egg-binding/Premature Egg Release

Livebearer females produce eggs, just like any other fish. In Poeciliids, fertilised eggs are retained internally until they hatch, just prior to birth. Occasionally, something goes wrong somewhere along the line, with the affected female simply getting progressively fatter and being unable, eventually, to give birth. Old females nearing the end of their reproductive life seem to be more prone to this problem than younger, healthier ones. When egg-binding occurs, death is virtually inevitable.

Premature egg-release also happens from time to time in Poeciliid live-bearers. In these cases, partly-developed embryos, still with a larger or smaller part of their yolk supply unconsumed, and still bearing their egg membranes, are ejected by the affected female.

If premature egg release occurs relatively late during gestation, the embryos are able to complete their development successfully, provided that water and other conditions are particularly good.

However, since chilling sometimes appears to be linked with premature egg release, especially in females that are netted, and therefore stressed, during the later stages of gestation, the chances of rearing such embryos to full maturity are generally quite slim.

Siamese Twins

In some ways, this is a misleading term because, to most people, Siamese twins are usually of equal, or nearly equal, size to each other.

This may, indeed, often be so in livebearers (usually Guppies), but on numerous other occasions the minor twin (referred to as a teratoma) appears as no more than an appendage or deformed blob attached to the

Plate 32   Siamese Twins, such as this Green Molly pair, are not too uncommon among intensely selected cultivated stocks of livebearers. (*Ross Socolof*)

vent of the dominant twin. Where the size discrepancy is as large as this, the 'pair' will often grow to maturity. Where both twins are more or less equal, long-term survival seems to be more difficult.

Siamese twinning has, to the best of my knoweledge, only been reported in cultivated varieties of livebearers. This does not, of course, eliminate the possibility of twinning occurring in wild-type livebearers – but, if it does, it does not appear to have been recorded.

Humpback

Some types of livebearers – cultivated varieties of Guppy in particular – exhibit a tendency towards this condition.

While an inadequate diet may be partly responsible in some cases, this progressive (and usually irreversible) affliction usually affects older individuals. The most highly susceptible are old females, which may appear to be perfectly alright as they enter their last gestation, but emerge slightly humpbacked once they have given birth. They may survive for many months in this condition, apparently not experiencing pain, but will neither regain their normal shape (in fact, it becomes more accentuated with time) nor produce further fry.

Cancerous Pigmentation

Some types of Poeciliids – usually, but not exclusively, cultivated varieties – are susceptible to tumour formation. The best-known examples occur among Swordtail × Platy hybrids, where melanomas (black-pigmented skin tumours) are not particularly uncommon. The condition is untreatable but not infectious.

A similar skin condition is known in *Limia vittata* (the Cuban Limia) in which the normal yellow-pigmented cells that this species possesses may proliferate to the extent that they may develop into xanthomas – yellow versions of the black melanoma. There seems to be some doubt as to whether this condition is truly cancerous or merely pre-cancerous.

Other Diseases

These include the so-called non-specific diseases, in which the symptoms can be anything from shimmying (swimming on the spot, with pronounced body movements and clamped fins), to loss of appetite, loss of colour, exophthalmia (Pop-eye), abnormal behaviour, and a whole host of other signals which indicate that the fish is not well, but which do not always allow for accurate diagnosis of the causative agent.

Also included are the common diseases like White Spot, or Ich, and the not-so-common ones like Roundworm (Nematode) infections.

In the case of non-specific diseases, the symptoms can usually be taken to indicate that environmental conditions are unfavourable in one way or another. Therefore, the first step should consist of a series of investigations of the aquarium environment, including water tests. The most common of these environmentally influenced problems, their sources and suggested remedial steps are summarised in the accompanying chart.

PROBLEMS CAUSED BY ADVERSE ENVIRONMENTAL CONDITIONS

Causative Agent	Some Symptoms	Some Preventative Steps and Remedies
Chlorine	Restless movements; loss of balance	Use dechlorinator or vigorously aerate water for 24 hours
Chloramine and Ammonia	Inflamed gills and fin edges; blood spots; loss of balance	Use dechlorinators, accompanied by ammonia-absorbing medium such as zeolite or activated charcoal (some dechloraminators now available). Hydrogen peroxide and sodium thiosulphate may also be used
Nitrites and Nitrates	Similar to Ammonia symptoms	Partial water changes. Use of de-nitrifying agent (for nitrate reduction). Establish good biological purification system (for nitrite reduction)

Causative Agent	Some Symptoms	Some Preventative Steps and Remedies
Metallic Ions	Inflamed, clumped gill filaments; accelerated respiratory movements; gasping at surface	Allow tap to run for several minutes before water is drawn for an aquarium. Use tapwater conditioners. Carry out immediate partial water change if poisoning is suspected
Oxygen and Nitrogen	(i) *Excess:* Gas Bubble Disease: small bubbles visible under skin, in fins and around head and eyes; Exophthalmia (Pop-eye)	(i) Locate tank away from direct sunlight if overstocked with plants. Avoid sudden cutting off of light in well-oxygenated, well-planted tanks. Avoid major changes/replacement of well-aerated aquarium water with poorly oxygenated tapwater
	(ii) *Insufficient oxygen:* Gasping at surface; some loss of colour	(ii) Initiate vigorous aeration. Carry out partial water change. Spray water on surface
pH	(i) *Acidosis:* (a) Fast swimming movements; gasping at surface; occasional jumps out of the water; or (b) Extreme sluggishness; tendency to hide; loss of colour and appetite	(i) Avoid overstocking with fish and understocking with plants. Carry out immediate partial water change. Add appropriate proprietary pH adjuster
	(ii) *Alkalosis:* Serious damage to gills; disintegration of fin edges; general 'opaqueness' of skin	(ii) Locate heavily planted tanks away from prolonged, direct sunlight. Carry out immediate partial water change. Add appropriate proprietary pH adjuster
Fumes and Sprays	Generally as for metallic ions	Carry out immediate partial water change. Ventilate room. Switch off aeration until level of fumes has subsided

Causative Agent	Some Symptoms	Some Preventative Steps and Remedies
Temperature	(i) *Low:* Shimmying; sluggish movements; resting on bottom; reduced gill cover, fin and body movements; some loss of coloration	(i) Check heater and thermostat. Increase temperature gradually by replacing amounts of aquarium water with warmer water in a series of *small* water changes. Switch on heater-thermostat
	(ii) *High:* Initial intensification of coloration; increased level of activity (above normal); increased rate of respiration; gasping at surface	(ii) Check heater and thermostat. Switch off heater-thermostat until temperature has gradually dropped to required level. Carry out a series of *small* water changes to reduce temperature slowly. Ice cubes may effect a slight drop

NOTE: Text of chart based largely on *The Tropical Freshwater Aquarium* by John Dawes, published by Hamlyn (1986). ISBN: 0–600–30649–6.

Diseases caused by pathogenic organisms can be difficult to diagnose and treat successfully without specialised knowledge. I have, therefore, included several comprehensive works on the subject under Further Reading at the end of this chapter and would urge the interested reader to consult one or other of the listed works for detailed information.

Further Reading

Allgayer, Robert and Teton, Jacques, *The Complete Book of Aquarium Plants*, Ward Lock, p. 157 (1987).

Andrews, Chris, Exell, Adrian and Carrington, Neville, *The Interpet Manual of Fish Health*, Salamander Books, p. 208 (1988).

Cowie, C. B., Mackie, A. M. and Bell, J. G. (eds), *Nutrition and Feeding in Fish*, Academic Press, p. 489 (1985).

Dawes, John, *The Tropical Freshwater Aquarium*, Hamlyn, p. 160 (1986).

Dawes, John, 'Live Fry Food Menu', *Aquarist & Pondkeeper*, 53 (8) pp. 29–30 (November 1988).

Dawes, John, *Bolstering Sales of Brackish Water Fish*, Pets Supplies Marketing, pp. 21–27 (July 1989).

Ellis, Anthony E. (ed), *Fish and Shellfish Pathology*, Academic Press, p. 412 (1985).

Goldstein, Robert, 'Live Foods', *Aquarist & Pondkeeper*, 52 (1), pp. 43–45 (April 1987).

Gordon, Myron and Axelrod, Herbert R., *Fancy Swordtails*, Tropical Fish Hobbyist Publications, Inc., p. 96 (1968).

Goss, Michael W., *Brackish Aquariums*, Tropical Fish Hobbyist Publications, Inc., p. 93 (1979).

James, Barry, *A Fishkeeper's Guide to Aquarium Plants*, Salamander Books, p. 118 (1986).

Masters, Charles O., *Encyclopedia of Live Foods*, Tropical Fish Hobbyist Publications, Inc., p. 336 (1975).

Mühlberg, Helmut, *The Complete Guide to Water Plants*, E. P. Publishing, p. 392 (1981).

Post, George, *Textbook of Fish Health*, Tropical Fish Hobbyist Publications, Inc., p. 288 (1987).

Scott, Peter W., *A Fishkeeper's Guide to Livebearing Fishes*, Salamander Books, p. 118 (1987).

Untergasser, Dieter, *Handbook of Fish Diseases*, Tropical Fish Hobbyist Publications, Inc., p. 160 (1989).

Part III
Selected Species
and Varieties

Catalogue of Selected Species and Varieties

Introduction

No book on livebearers, however exhaustive it may claim to be, can pretend to illustrate and describe every known living species and variety. Photographs may be not readily available or even not available at all. Then, the species or variety may be so rarely seen, or have been kept by so few people, that it would seem more appropriate to allocate space to other fish. Better-known, more common or relatively uncommon species that are likely to become more readily available in the foreseeable future have therefore been given priority in this book.

Fish taxonomy and nomenclature are two fields of study that are constantly resulting in ever-more-accurate refinements and additions to our already considerable list of species. For instance, research currently being carried out on the Goodeid fish we know as *Ataeniobius toweri* may well result in the allocation of this species to the genus *Goodea*, with the name *Ataeniobius* ceasing to be regarded as valid. Had this book been written a few years ago, then *Characodon audax* would have been referred to simply as the Black Prince (a most appropriate name coined by the British aquarist Ivan Dibble). Similarly, the various *Limia* species would have appeared as *Poecilia* species, *Girardinichthys viviparus* would have appeared as *Limnurgus innominatus* and the question of regarding *Gambusia affinis* and *Gambusia holbrooki* as two separate species would not even have arisen.

While the majority of generic and specific names in the pages that follow are likely to hold good for many years to come, some will, inevitably, undergo changes, a few probably doing so even as the book goes to press (as in the case of some *Xiphophorus* – Swordtail – species currently under review). I have made every effort to be as up to date as possible, but I would ask readers to bear this caveat in mind.

I have largely adopted Lynne Parenti's classification of livebearers at family and subfamily levels.

Selected Species and Varieties

FAMILY: **POECILIIDAE**
SUBFAMILY: POECILIINAE
SUPERTRIBE: POECILIINI
TRIBE: POECILIINI

GENUS: *Alfaro*: The most easily identifiable feature of this genus is a keel-like series of scales that extends along the ventral edge of the caudal peduncle. This characteristic has led to the common name of Knife Livebearers for the two species in the genus.

1. *Alfaro cultratus*
SYNONYMS: *Petalosoma cultratum*, *Alfaro acutiventralis*, *Alfaro amazonum* and others
COMMON NAME; Knife Livebearer
RANGE: Costa Rica, Panama, Nicaragua
OVERALL SIZE: Males around 9 cm (3.5 in); females 10 cm (4 in) – both usually smaller
WATER REQUIREMENTS: Clear moving water preferred. Temperature around 25 °C (77 °F) for general

Plate 33 *Alfaro cultratus* male. (*Bill Tomey*)

maintenance; up to 28 °C (82 °F) for breeding. Well-planted aquarium is advisable. Sometimes seems to benefit from a small amount of salt in the water, but this is not essential

PREFERRED DIET: Prefers swimming live-foods but will also take flaked foods

BREEDING: Broods of up to 100 fry produced every 4–5 weeks

NOTES: This is a shy, often nervous, and sometimes aggressive, species best suited for a species tank

2. *Alfaro huberi*

SYNONYMS: *Furcipenis huberi*, *Priapichthys huberi*

COMMON NAMES: Orange Rocket, Orange Knife Livebearer

RANGE: Southern Guatemala to Nicaragua

OVERALL SIZE: Early reports indicated males 3.5–5 cm (1.4–2 in); females 7 cm (2.8 in). Recent data show that this species is at least as large as *A. cultratus*

WATER REQUIREMENTS AND DIET: As for *A. cultratus*

BREEDING: Early data indicated smaller broods than *A. cultratus* (perhaps up to 80). More recent accounts report up to 100 fry

NOTES: If anything, this fish is more nervous than *A. cultratus* and more sensitive to poor water conditions. Can take up to one year to mature

GENUS: *Limia*: This genus was synonymised with *Poecilia* by Rosen and Bailey in 1963. It was, however, re-allocated to full generic status by Luis Rivas in 1978. ('A New Species of Poeciliid Fish of the Genus *Poecilia* from Hispaniola, with Reinstatement and Redescription of *P. dominicensis* (Evermann and Clark)' in *Northeast Gulf Science*, Vol. 2, No. 2, pp 98–112, December 1978.) He based his conclusions on: (i) Gonopodial characteristics found in *Limia* species but not in *Poecilia*. For example, 'subdistal segments of ray 3 of gonopodium without spines or processes' = *Limia*; '... with erect or retrorse spines' = *Poecilia*; and (ii) Distribution: Confined to Greater Antilles, i.e. Cuba, Hispaniola (Haiti and Dominican Republic), Jamaica and Grand Cayman Island = *Limia*; mainland of North, Central and South America, and the island of Hispaniola = *Poecilia*.

Plate 34   *Alfaro huberi* male. (*Dennis Barrett*)

Parenti and Rauchenberger follow Rosen and Bailey's classification. Pending full revision of the genus *Limia* (with two subgenera, *Limia* and *Odontolimia*), I have adopted Rivas' nomenclature, basing my decision largely on gonopodial features. This classification is also accepted by, among others, Meyer, Wischnath and Foerster (*Lebendge-bärende Zierfische*, published by Mergus, 1985)

Subgenus *Limia*

1. *Limia caymanensis*
COMMON NAME: Grand Cayman Limia
RANGE: Grand Cayman Island
OVERALL SIZE: Males around 3 cm (1.2 in); females 3.5–4.5 cm (1.4–1.8 in) – probably grow larger than this
WATER REQUIREMENTS: Neutral or slightly alkaline water at around 25 °C (77 °F)
PREFERRED DIET: Live and dry foods (latter incorporating a vegetable component)

BREEDING: Smallish broods reported so far (around 20–25 fry). Not yet bred regularly in aquaria
NOTES: This is one of the more recent *Limia* species to be described (Rivas and Fink, 1979). It has not yet been kept with any regularity by aquarists

2. *Limia dominicensis*
SYNONYM: *Poecilia dominicensis, Limia tridens*
COMMON NAME: Tiburón Peninsula Limia (name hardly ever used)
RANGE: Hispaniola (Haiti)
OVERALL SIZE: Male around 3 cm (1.2 in); females 4–4.5 cm (1.6–1.8 in) – sometimes slightly larger
WATER REQUIREMENTS: Slightly alkaline water at around 25 °C (77 °F) has been found quite satisfactory
PREFERRED DIET: As for *B. caymanensis*
BREEDING: Broods of around 30 fry every 6 weeks or so (but larger broods possible)
NOTES: This is not a particularly hardy species, but is very attractively marked when in peak condition

Plate 35 *Limia caymanensis* male. (*Dennis Barrett*)

Plate 36 *Limia dominicensis* pair (male below), from Port au Prince, Haiti. (*Manfred Meyer*)

3. *Limia melanogaster*
SYNONYMS: *Poecilia melanogaster, Limia caudofasciata tricolor, Limia tricolor*
COMMON NAMES: Blue Limia, Black-bellied Limia
RANGE: Jamaica
OVERALL SIZE: Males up to 4.5 cm (1.8 in); females around 6 cm (2.4 in)
WATER REQUIREMENTS: Chemical composition not critical. Temperatures between 22–26 °C (72–79 °F) are quite adequate
PREFERRED DIET: As for *L. caymanensis*
BREEDING: Smallish broods of around 25 fry produced every 6 weeks or so
NOTES: *L. melanogaster* males are sexually very active and will constantly harass females. Males of this species will hybridise quite easily with females of other *Limia* species

4. *Limia nigrofasciata*
SYNONYMS: *Poecilia nigrofasciata, Limia arnoldi*
COMMON NAMES: Humpbacked or Black-barred Limia
RANGE: Hispaniola (Haiti)
OVERALL SIZE: Males up to 5.5 cm (2.2 in); females around 6 cm (2.5 in)
WATER REQUIREMENTS: Neutral or slightly alkaline water kept at around 26 °C (79 °F) seems to suit this species best
PREFERRED DIET: As for *L. caymanensis*
BREEDING: Larger broods than most other *Limia* species – around 50 fry every 6–10 weeks
NOTES: Temperature appears to affect the male:female ratio of offspring (see Sex Determination in Part II). In addition to the characteristic hump, *L. nigrofasciata* males also have a pronounced 'keel' between the anal and caudal fins

5. *Limia perugiae*
SYNONYMS: *Poecilia melanonotata, Platypoecilus perugiae*
COMMON NAME: Perugia's Limia
RANGE: Hispaniola – Rosen and Bailey report *L. perugiae* from the Dominican Republic and *L. melanonotata* from Haiti. Despite recognition of the synonymy, Meyer, Wischnath aad Foerster report *B. perugiae* as coming from the Dominican Republic but make no mention of the Haitian *P. melanonotata* population
OVERALL SIZE: Males 5–7 cm (2–2.8 in); females 5–8.5 cm (2–3.3 in)
WATER REQUIREMENTS, DIET AND BREEDING: Generally as for *L. nigrofasciata*
NOTES: Jacobs (1969) states that *Poecilia (Limia) melanonotata* appears to be related to *L. nigrofasciata*. In view of the *L. melanonotata/perugiae* synonymy, this statement should perhaps be treated with some caution since *L. perugiae* looks very different from *L. nigrofasciata*

6. *Limia vittata*
SYNONYMS: *Poecilia vittata, P. cubensis, Gambusia vittata, G. cubensis, Limia cubensis, L. pavonina*
COMMON NAME: Cuban Limia
RANGE: Cuba
OVERALL SIZE: Males 5–6.5 cm (2–2.5 in); females up to 12 cm (4.7 in)
WATER REQUIREMENTS AND DIET: Generally as for *L. nigrofasciata*. Addition of a small amount of salt has occasionally been recommended
BREEDING: Approximately 50 fry produced every 4–6 weeks
NOTES: Some specimens possess very attractive black/yellow speckling, while others are largely devoid of colour. Will hybridise with other species, particularly *L. melanogaster*

7. OTHER *Limia* SPECIES
Meyer, Wischnath and Foerster recognise the following species of *Limia* in addition to the above. The common names are drawn from Gary K. Meffe's 'List of Accepted Common Names of Poeciliid Fishes' in *Ecology and Evolution of Livebearing Fishes* (Simon & Schuster, pp 450, 1989)

Plate 37 *Limia melanogaster* – two males and one female. (*Bill Tomey*)

Plate 38 *Limia nigrofasciata* pair (male above), from Lake Miragoane, Haiti. (*Manfred Meyer*)

Plate 39 *Limia perugiae* male. (*Dennis Barrett*)

Plate 40 *Limia vittata* male. (*John Dawes*)

Subgenus *Odontolimia*

L. fuscomaculata (Blotched Limia); Haiti

L. garnieri (Garner's Limia); Haiti

L. grossidens (Largetooth Limia); Haiti

L. immaculata (Plain Limia); Haiti

L. miragoanensis (Miragoane Limia); Haiti

L. ornata (Ornate Limia); Haiti

Subgenus *Limia*

L. pauciradiata (Few-rayed Limia); Haiti

L. sulphurophila (Sulphur Limia); Dominican Republic

L. versicolor (Vari-coloured Limia); Haiti

L. yaguajali (Yaguajal Limia); Haiti and Dominican Republic

L. zonata (Striped Limia); Dominican Republic

Neutral to slightly alkaline water kept at around 25 °C (77 °F), plus a varied diet incorporating a vegetable component, should provide adquate conditions for most of these species

GENUS: *Poecilia*: This is one of the two largest genera in the subfamily Poeciliinae (the other being *Xiphophorus*). *Poecilia* males have a distinct fleshy palp around the tip of the gonopodium (lacking in *P. heterandria* and very small in *P. elegans*). The pelvic fins of males are also modified in such a way that they aid in stabilising and directing the swing of the gonopodium during copulation

The following subgenera have been recognised: *Lebistes*, *Limia*, *Odontolimia*, *Pamphorichthys*, and *Poecilia*

Rivas afforded full species status to *Limia* and subgeneric status (within *Limia*) to *Odontolimia* and this is the approach I have adopted here

A. Subgenus: *Lebistes*

1. *Poecilia branneri*
SYNONYMS: *Micropoecilia branneri*, *Poecilia heteristia*
RANGE: Pará in Brazil – believed to be the southernmost representative of the subgenus
OVERALL SIZE: Males about 2.5 cm (1 in); females about 3–3.5 cm (1.2–1.4 in)
WATER REQUIREMENTS: Densely planted brackish water (approx. 5% salt) is recommended. Temperature 24–28 °C (75–82 °F)
PREFERRED DIET: Small foods, including both a live and a vegetable component
BREEDING: This has proved to be a difficult species to breed, with small numbers of fry (maximum reported is 19) produced at frequent intervals
NOTES: *P. branneri* is an extremely attractive, rarely seen species. The recent introduction of commercially available rotifer hatching/rearing kits could well improve the chances of rearing fry successfully, thus helping to increase the popularity of this delightful fish

2. *Poecilia picta*
SYNONYMS: *Micropoecilia picta*, *Acanthophacelus melanzonus*
COMMON NAME: Black-banded Poecilia
RANGE: Demerara River in British Guyana; Trinidad; Brazil
OVERALL SIZE: Males up to 3 cm (1.2 in); females up to 5 cm (2 in)
WATER REQUIREMENTS AND DIET: Generally as for *P. branneri*
BREEDING: Perhaps slightly easier than *P. branneri* although broods are of a similar size
NOTES: This is a beautiful Guppy-like species that is slowly becoming more widely available

Plate 41 *Poecilia branneri* male, from Belem, Brazil. (*Manfred Meyer*)

Plate 42 *Poecilia picta* male. (*Dennis Barrett*)

Plate 43 *Poecilia reticulata* (Guppy) – wild-type male from Venezuela. (*Manfred Meyer*)

3. *Poecilia reticulata*

SYNONYMS: *Lebistes reticulatus, Girardinus reticulatus, Acanthophacelus guppii, Heterandria guppii, Poecilia poeciloides* and others

COMMON NAMES: Guppy, Millions Fish

RANGE: Widely distributed north of the Amazon: Dutch Antilles, Trinidad, Windward Islands, Barbados, Grenada, Antigua, Leeward Islands, St Thomas, Venezuela and British Guyana. Introduced into numerous exotic locations (see Table IX for details)

OVERALL SIZE: Males around 3 cm (1.2 in); females around 5 cm (2 in) – cultivated varieties are considerably larger

WATER REQUIREMENTS: Wide range of conditions tolerated in terms of both temperature and chemical composition. Appropriate temperature range for aquarium maintenance and breeding: 21–25 °C (70–77 °F). A small amount of salt – 5 ml (1 teaspoonful) per 4.5 litres (1 Imperial gallon) – may be found beneficial at least for wild-caught specimens

PREFERRED DIET: Will eat a wide range of small live, frozen, freeze-dried and dry foods with a regular vegetable component

BREEDING: Very prolific species producing broods every 4–6 weeks. As many as 193 fry have been recorded from a single female by an aquarist (Joseph Camin) but this is most atypical

NOTES: After *Gambusia affinis* and *G. holbrooki*, the Guppy is the most widely distributed Poeciliine species. Numerous varieties exist in the wild, while even more numerous body and fin configurations have been developed by commercial breeders and hobbyists worldwide, making the Guppy one of the most popular fish in the history of the aquarium hobby. A fairly recent introduction is Endler's Livebearer, an exceptionally beautiful true-breeding, short-finned variety much admired by specialists

4. OTHER MEMBERS OF THE SUBGENUS *Lebistes*

Three other species of this subgenus are encountered from time to time:

Poecilia amazonica Pará, Brazil
Poecilia parae British Guyana; Pará, Brazil
Poecilia scalpridens Pará, Villa Bella and Matto Grosso in Brazil

B. Subgenus: *Poecilia*

5. *Poecilia butleri*

SYNONYM: *Platypoecilus nelsoni*

COMMON NAME: Pacific Molly

RANGE: Pacific coast of Mexico and Panama

OVERALL SIZE: Males around 7 cm (2.8 in); females around 8 cm (3.2 in)

WATER REQUIREMENTS: Slightly alkaline, hardish water with some salt – 5 ml (1 teaspooful) per 4.5 litres (1 Imperial gallon). Temperature around 24–28 °C (75–82 °F)

PREFERRED DIET: Live, flake, freeze-dried and frozen foods, plus a regular vegetable component

BREEDING: Large broods are obviously possible but most successful breeding attempts appear to have produced only some 20 fry

NOTES: This species is very similar to *P. sphenops* and is, in fact, regarded as such by some workers

6. *Poecilia caucana*

SYNONYMS: *Girardinus caucanus, Mollienisia caucana, Allopoecilia caucana*

RANGE: Río Cauca, Antioquia in Colombia; Pacific drainages, Darién in Panama and Lago de Maracaibo in Venezuela

OVERALL SIZE: Males up to 3.5 cm (1.4 in); females around 5 cm (2 in)

WATER REQUIREMENTS: Well-planted, slightly alkaline and slightly hard water. Salt – 5 ml (1 teaspoonful) per 4.5 litres (1 Imperial gallon) – optional. Temperature: 20–27 °C (68–80 °F)

Plate 44 *Poecilia reticulata* (Guppy) – Blue Star male produced by Providence Tropicals in Florida. (*Harry Grier/Florida Tropical Fish Farms Association*)

Plate 45 *Poecilia butleri* male, from Río Balsas, Mexico. (*Manfred Meyer*)

Plate 46   *Poecilia caucana* male, from Tolu, Colombia. (*Manfred Meyer*)

PREFERRED DIET: Wide range of small foods, which should include a vegetable component

BREEDING: Although some of the literature indicates up to 100 fry every 4–8 weeks, highest numbers actually recorded are around the 80 mark

NOTES: This is a peaceful, attractive species with a relatively low level of fry cannibalism if the adult fish are well fed

7.   *Poecilia chica*

COMMON NAME: Dwarf Molly

RANGE: Ríos Cuetzmala, Purificación and Cihualtán in Jalisco, Mexico

OVERALL SIZE: Males around 3 cm (1.2 in); females about 3.5 cm (1.4 in)

WATER REQUIREMENTS: Neutral to slightly alkaline water of medium hardness. Temperature tolerance very wide – between 21–32 °C (70–90 °F) reported

PREFERRED DIET: Primarily vegetable-based but will also accept flake and other foods

BREEDING: About 30 fry produced every 5 weeks or so

NOTES: A very small dorsal fin is typical of this species (*chica* means small in Spanish). Hybridisations have been recorded between this species and Lyretail Black Mollies by Norbert Dokoupil (see *Tropical Fish Hobyist*, September 1987, pp 47–51). The hybrids proved fertile

8.   *Poecilia 'formosa'*

COMMON NAME: Amazon Molly

RANGE: Southeastern Texas and northeastern Mexico in conjunction with *P. latipinna* in coastal waters and *P. mexicana* in inland rivers

OVERALL SIZE: Approximately 5–6 cm (2–2.4 in). No males exist

WATER REQUIREMENTS: Coastal populations: hard alkaline water with 5 or 10 ml (1 or 2 teaspoonfuls) of salt per 4.5 litres (1 Imperial gallon). Inland populations do not require the salt. Temperature 24–27 °C (75–80 °F)

PREFERRED DIET: Wide range of food, but must include a vegetable component

BREEDING: Not particularly prolific in aquaria. Broods, of around 30 fry, are produced sporadically

NOTES: This is not a true species but a natural hybrid involving *P. latipinna* and *P. mexicana*. Copulation is necessary for reproduction and a variety of males from other *Poecilia* species have been successfully used for this purpose. From time to time, male *P. 'formosa'* are reported in the hobby but none have been validated with any certainty

Plate 47   *Poecilia chica* male. (*Wilf Blundell*)

Plate 48   *Poecilia 'formosa'* – the Amazon Molly. (*Terry Waller*)

Plate 49 *Poecilia latipinna* (male) – one of the Sailfin Mollies. Not all males exhibit the large sail-like dorsal fin. This specimen came from a Texas population, USA. (*Manfred Meyer*)

9. *Poecilia latipinna*

SYNONYMS: *Mollienesia latipinna, Poecilia multilineata, Limia poeciloides, Gambusia poeciloides* and others

COMMON NAME: Sailfin Molly

RANGE: South and North Carolina, Virginia, Texas, Florida, Atlantic coast of Mexico

OVERALL SIZE: Males around 10 cm (4 in); females around 12 cm (4.7 in)

WATER REQUIREMENTS: Slightly hard alkaline water containing about 5 ml (1 teaspoonful) of salt per 4.5 litres (1 Imperial gallon). Temperature around 25–28 °C (77–82 °F)

PREFERRED DIET: Will accept a wide range of foods but needs a regular vegetable component for long-term health

BREEDING: Very large broods of around 130–140 are not uncommon. Gestation period up to 8–10 weeks

NOTES: Not all males produce the characteristic sail-like dorsal fin. Colour variations occur even in the wild. In captivity, *P. latipinna* has been cross-bred with other species and varieties and has contributed significantly to the wide range of aquarium Mollies currently available. *P. latipinna* is also one of the ancestral species linked with *P. 'formosa'*

10. *Poecilia mexicana*

SYNONYMS: *Poecilia thermalis, P. chisoyensis, Mollienesia sphenops vantynei* and others

COMMON NAME: Atlantic Molly

RANGE: Northern Mexico, Guatemala, Honduras

OVERALL SIZE: Males up to 7 cm (2.8 in); females up to 8.5 cm (3.4 in)

WATER REQUIREMENTS: Alkaline, medium-hard conditions preferred. Temperature 24–27 °C (75–80 °F)

PREFERRED DIET: Range of foods accepted. Vegetable component should be included

BREEDING: Most reports quote 30 or so fry but some go as high as 75–80. Gestation 4–6 weeks

NOTES: This fish is very sphenops-like in overall appearance and is believed by some to be the ancestral species from which many aquarium Mollies have been developed. Most references quote *P. sphenops* as the ancestral species but this can be explained by the fact that *P. mexicana* was long regarded as *P. sphenops*. A cave morph of *P. mexicana* has been collected by Ross Socolof. This collection included a range of characteristics from eyeless, somewhat deformed individuals to fully eyed ones. Along with *P. latipinna*, *P. mexicana* is one of the species involved in the *P. 'formosa'* complex (see Sex Determination in Part II for further details)

11. *Poecilia petenensis*

SYNONYM: *Mollienesia petenensis*

COMMON NAMES: Spiketail Molly, Petén Molly

RANGE: Río Usumacinta drainage and surrounding lakes in Petén, Guatemala; also found in Tabasco, Quintana Roo and Chiapas in Mexico, and in Belize

OVERALL SIZE: Males 8–13 cm (3.2–5.1 in); females 6–10 cm (2.4–4 in)

Plate 50 *Poecilia mexicana* male, from the Río Panuco in Mexico. (*Manfred Meyer*)

Plate 51 *Poecilia mexicana* (cave morph) – reportedly, many of the specimens collected from the wild show this characteristic deformity of the spine. (*Ross Socolof*)

WATER REQUIREMENTS: Alkaline, medium-hard water, preferably with 5–10% seawater/salt added. Temperature 23–28 °C (73–82 °F)

PREFERRED DIET: As for *P. mexicana*

BREEDING: Broods of around 100 fry produced every 5–7 weeks or so

NOTES: This species is very impressive, particularly when males are displaying to each other. The sail-like dorsal, while not as spectacular as in *P. velifera*, is nevertheless bigger than in most *P. latipinna*. Males are also characterised by a very short sword in the caudal fin

12. *Poecilia sphenops**

SYNONYMS: *Mollienisia sphenops*, *Poecilia gillii*, *P. dovii*, *P. spilurus*, *P. mexicana*, *P. vandepolli*, *Gambusia plumbea*, *Platypoecilus mentalis* and many others

COMMON NAMES: Green Molly, Sphenops Molly, Pointed-mouth Molly, Mexican Molly, Liberty Molly**

RANGE: From Texas down through Mexico, along both coasts to Colombia but also introduced elsewhere (see Table IX)

OVERALL SIZE: Males around 6 cm (2.4 in); females around 8 cm (3.2 in)

WATER REQUIREMENTS AND DIET: As for *P. petenensis*

BREEDING: Fairly large broods of around 80 fry every 5–7 weeks. Larger broods have been recorded from aquarium stocks (but see below)

NOTES: This is a very variable species, a factor that often makes precise identification difficult. Many other species have been synonymised with it at one time or other, only to be accorded full specific status later on. Further refinements will undoubtedly follow in the future

* *P. sphenops* is usually quoted as the ancestral species to many of the shorter-finned aquarium varieties of Molly. However, a growing body of opinion now believes that the correct ancestral species is *P. mexicana*, the confusion arising out of these two species having being regarded as one and the same in the past

** The Liberty Molly is a variety of *P. sphenops* which possesses a very attractively marked dorsal fin bearing numerous speckles on a predominantly red base, plus a reddish caudal fin

Plate 52 *Poecilia petenensis* (the Spiketail Molly). This mature male, collected for scientific purposes, has been laid out to show the large sail-like dorsal fin, the spike on the base of the caudal fin and the short gonopodium characteristic of the species. (*Ross Socolof*)

Plate 53   *Poecilia sphenops* (Green Molly) – an attractive male with an orange gonopodium, collected in the Río Panuco drainage in Mexico. (*Manfred Meyer*)

Plate 54   Gold Dust Molly – a short-finned aquarium variety produced by Summerland Tropical Fish Farm in Florida. (*Harry Grier/Florida Tropical Fish Farms Association*)

Plate 55 A wild-caught *Poecilia velifera* (Sailfin) male, from Progresso in Mexico. (*Manfred Meyer*)

13. *Poecilia velifera*
SYNONYM: *Mollienisia velifera*
COMMON NAMES: Sailfin Molly, Yucatán Molly
RANGE: Yucatán Peninsula, Mexico
OVERALL SIZE: Males up to 15 cm (6 in); females up to 18 cm (7 in)
WATER REQUIREMENTS AND DIET: As for *P. petenensis*
BREEDING: Average broods of around 50 fry are produced every 6–8 weeks. Broods of more than 100 fry are not uncommon
NOTES: This is the largest of the Sailfin Mollies and, like them, is sensitive to poor water quality, developing shimmying if conditions are not right. Numerous commercially developed colour varieties exist

14. *Poecilia vivipara*
SYNONYMS: *Molinesia surinamensis*, *Mollienisia surinamensis*, *Poecilia surinamensis*, *P. unimaculata*, *P. unimacula*, *P. holacanthus*, *Neopoecilia holacanthus*
COMMON NAME: One-spot Livebearer
RANGE: Aruba, Curaçao, Bonaire and the Venezuelan islands; Trinidad; Leeward Islands northward to Martinique; western Venezuela and along the coast, through Brazil, to Río de la Plata in Argentina. Rosen and Bailey also report it as introduced into Puerto Rico
OVERALL SIZE: Males up to 6 cm (2.4 in); females up to 8.5 cm (3.5 in) – both are usually smaller than this in aquaria

Plate 56   *Poecilia velifera* – a spectacular Green Sailfin male. (*Bill Tomey*)

Plate 57   A superb Sunset Sailfin, from Blackwater Fishery in Florida. (*Harry Grier/Florida Tropical Fish Farms Association*)

Plate 58 A Black Lyretail Sailfin – but is it pure *P. velifera* or is there some *P. sphenops* and *P. latipinna* in it as well? This specimen was produced by Liles Tropical Fish Farm in Florida. (*Harry Grier/Florida Tropical Fish Farms Association*)

WATER REQUIREMENTS: Water conditions not generally critical but extremes of pH and hardness should be avoided. Wide temperature tolerance reported: 18–27 °C (65–80 °F) – I have collected this species (along with *Jenynsia*) in the highly polluted brackish Lago Rodrigo de Freitas in Rio de Janeiro in a water temperature of 34.4°C (94°F)! A little salt may benefit some wild specimens collected from brackish waters

PREFERRED DIET: Varied composition but including livefoods

BREEDING: Average broods of 50–60 fry every 4–8 weeks. A record spawning of 210 fry has been reported by an aquarist (C. L. Michaels)

NOTES: Not surprisingly, owing to its wide distribution, this is a variable species. For instance, the intensity of the characteristic body spot can range from spectacular to virtually non-existent

Plate 59   *Poecilia vivipara* female from Río de Janeiro, Brazil – showing the characteristic black body spot, which fades with age. (*Manfred Meyer*)

Plate 60   *Poecilia vivipara* male – the black body spot is quite faint in this mature aquarium specimen. (*Wilf Blundell*)

Plate 61   *Poecilia maylandi* male, from Río Chacamero in Mexico. This species is superficially like *P. sphenops*. (*Manfred Meyer*)

15. OTHER MEMBERS OF THE SUBGENUS *Poecilia*

In addition to the above, there are quite a few other species in this subgenus, the exact number varying according to the authority consulted. The following is a complete listing of these other species based on Meyer, Wischnarth and Foerster: *Lebendgebärende Zierfische* (Mergus, 1985). The common names are drawn from: 'List of Accepted Common Names of Poeciliid Fishes' by Gary K. Meffe in *Ecology and Evolution of Livebearing Fishes*, (Simon & Schuster, p 450 1989)

Poecilia catemaconis (Catemaco Molly) Laguna Catemaco in Veracruz, Mexico
Poecilia dominicensis Hispaniola (mainly highlands)
Poecilia elegans (Elegant Molly) Dominican Republic
Poecilia hispaniolana (Hispaniola Molly) Central Hispaniola
Poecilia latipunctata (Tamesí Molly) Tamaulipas, Mexico
Poecilia maylandi Guerrero, Mexico
Poecilia sulphuraria Tabasco, Mexico

16. MEMBERS OF THE SUBGENUS *Pamphorichthys*

Members of this subgenus are only rarely encountered within the aquarium hobby

Poecilia hasemani Puerto Suarez and Río Paraguay in Bolivia
Poecilia heterandria La Guaira in Venezuela
Poecilia hollandi Río Sao Francisco, Río Itapicurú, Baia and San Paolo in Brazil
Poecilia minor Villa Bella and Mato Grosso in Brazil

GENUS: *Priapella*: Dentition and gonopodial characteristics exhibited by this genus indicate that it has certain affinities with the genus *Xiphophorus* (Swordtails and Platies). Quoting from Rosen and Bailey (1963): '. . . ray 3 terminating in long bony hook that is followed by three to five oblong segments and a series of from 10 to 20 spines; thin crescent of tough membranous tissue dorsally on terminal hook of ray 3 that is similar to but smaller than blade in *Xiphophorus*'. In addition, in *Priapella* species: 'Pelvic fin (is) rather large but without terminal modification'. (In *Xiphophorus* there *are* modifications.)

1. *Priapella compressa*
RANGE: Palenque ruins in the Río Grijalva system, Chiapas, Mexico
OVERALL SIZE: Males 3.5–4.5 cm (1.4–1.8 in); females 6–7 cm (2.4–2.8 in)
WATER REQUIREMENTS: Clean, well-aerated water is essential. Temperature 22–26 °C (72–79 °F)
PREFERRED DIET: Varied with a predominance of aquatic livefoods (this species rarely feeds off the bottom)
BREEDING: Smallish brood of 30 or fewer produced sporadically – although gestation period is between 5–8 weeks
NOTES: This is a very attractive, though nervous, fish with an orange-based body and turquoise eyes. Females show very little sign of 'filling up' during gestation. Signs of imminent birth are therefore difficult to discern

2. *Priapella intermedia*
RANGE: Mainly upper reaches of the Río Coatzacoalcos system in Oaxaca, Mexico
OVERALL SIZE: Males up to 5 cm (2 in); females 7 cm (2.8 in)
WATER REQUIREMENTS, DIET AND BREEDING: Generally as for *P. compressa*
NOTES: *P. intermedia* is a slimmer fish than *P. compressa*. It is also (usually) less colourful, although the white

Plate 62 *Priapella compressa* male – showing the deep body characteristic of the species. (*Wilf Blundell*)

streaks along the top and bottom of the caudal fin may be more vividly marked, at least in some specimens

3. OTHER *Priapella* SPECIES

Rosen and Bailey, Jacobs and other writers refer to a third species, *Priapella bonita*, from Refugio and Motzorongo in the upper reaches of Río Tonto, Río Papaloapan system in Veracruz, Mexico. If this species does exist (and many doubt its existence, especially since no 'bonitas' have been seen in the hobby), and it is not merely an isolated population of *P. intermedia*, it should be a really beautiful fish (*bonita* means pretty or beautiful in Spanish)

Perhaps the true 'bonita' is a new *Priapella* – unnamed at the time of writing – collected in 1987 by Manfred Meyer in the Río de la Palma, Mexico. It has a sky-blue body and red fins – quite a spectacular combination. This fish is currently being bred in captivity, so we could see another beautiful species of live-

bearer becoming available before too long

GENUS: *Xiphophorus*: Traditionally, this genus has been subdivided into three groups or complexes:

1. Helleri Group, including the true Swordtails typified by *X. helleri*.
2. Montezumae Group, typified by *X. montezumae* and *X. cortezi*
3. Maculatus Group, including the Platies, e.g. *X. maculatus*, *X. variatus*, *X. couchianus*, *X. xiphidium*

In 1979, however, Donn Rosen reassessed these groups in 'Fishes from the Uplands and Intermontane Basins of Guatemala: Revisionary Studies and Comparative Geography' (*Bulletin of the American Museum of Natural History*, Volume 162, Article 5, 1979)

He concluded that there were, indeed, three groups, but that their composition was as follows (he did not allocate overall names to these

Plate 63 *Priapella intermedia* is considerably slimmer than its close relative, *P. compressa. (Bill Tomey)*

groups, so I have added them for convenience):

1. Couchianus Group, consisting of *X. couchianus*, *X. meyeri* and *X. gordoni*, distinguished from the other groups in that the spines of gonopodial ray 3 are divided distally, while ray 4p carries distal serrae that converge at their tips
2. Maculatus Group, including *X. maculatus*, *X. variatus*, *X. evelynae*, *X. milleri* and *X. xiphidium*, i.e. the true Platies. The Platies are characterised by having the gonopodial blade bluntly rounded (as in *X. couchianus* and *X. gordoni*). They do not, however, possess the divided spines. Platies possess genes for sword production (demonstrated experimentally) but, except in the case of *X. xiphidium*, do not normally exhibit this character
3. Helleri Group, containing the true Swordtails, i.e. all the species that exhibit a shorter or longer sword.

Some, like *X. montezumae*, *X. pygmaeus* and *X. clemenciae*, have blunt blades like the Platies – as opposed to the pointed blade possessed by most Swordtail species – but are distnguished from them by the possession of the sword (very short in *X. pygmaeus*) and in having the tip of gonopodial ray 5a differentiated into a distinct claw-like segment. The Helleri Group is further subdivided into:

a) the *X. pygmaeus* and *X. nigrensis* subgroup
b) the *X. montezumae* and *X. cortezi* subgroup
c) the *X. clemenciae* subgroup
d) the *X. alvarezi* subgroup
e) the *X. helleri* subgroup
f) the *X. signum* subgroup

At the time of writing, a major review of some Swordtails is under way. Once this is published, the above classification is likely to undergo several significant changes

GROUP 1 (Couchianus Group)

1. *Xiphophorus couchianus*
SYNONYMS: *Limia couchiana, Gambusia couchiana, Mollienesia couchiana, Poecilia couchiana, Platypoecilus couchianus, X. couchianus couchianus*
COMMON NAMES: Monterrey Platyfish, Northern Platyfish
RANGE: Reportedly restricted to a spring and its outlet near Apodaca (a suburb of Monterrey) and to Mezquital (a lagoon and its outlet), all situated in Nuevo León in Mexico
OVERALL SIZE: Males approximately 3 cm (1.2 in); females up to around 6 cm (2.4 in), but generally smaller
WATER REQUIREMENTS: Well-planted aquaria with neutral or slightly alkaline water kept between 20–26 °C (68–79 °F)
PREFERRED DIET: Variety of small live and dry foods taken; diet should include a vegetable component
BREEDING: Small broods of around 20 fry produced sporadically
NOTES: This species is known to be under serious threat in the wild with, perhaps, only two natural popu-

lations remaining. Fortunately, there are various aquarium populations in the UK, West Germany and the US, so its future seems to be more promising than its status in nature would indicate. An *X. couchianus* × *variatus* hybrid referred to as *X. 'roseni'* is reported by Meyer, Wischnath and Foerster

A second, closely related species, *X. gordoni*, the Cuatro Ciénegas Platyfish (previously regarded as a subspecies of *X. couchianus*), is also highly threatened in its restricted natural distribution around the southeastern branch of the basin in Coahuila, Mexico. Aquarium stocks do exist but they are not as large as for *X. couchianus*

2. *Xiphophorus meyeri*
SYNONYMS: *Xiphophorus 'muzquiz', X. marmoratus*
COMMON NAME: Musquiz Platyfish
RANGE: Two communicating ponds near Muzquiz in Coahuila, Mexico
OVERALL SIZE: Probably as for *X. couchianus**
WATER REQUIREMENTS, DIET AND BREEDING: Probably as for *X. couchianus**

Plate 64 *Xiphophorus couchianus* – the Monterrey or Northern Platyfish. This male is from the Huasteca Canyon population in Nuevo León in Mexico. (*Manfred Meyer*)

Plate 65   A pair (male above) of Musquiz Platyfish (*Xiphophorus meyeri*) – collected from one of the only two ponds where this fish is found near Musquiz in Coahuila Mexico. (*Manfred Meyer*)

NOTES: This is the newest Platy species to date. It was first described by M. Schartl and J. H. Schroeder in 1988: *A new species of the genus* Xiphophorus (Senckenbergiana biol., 68, pp 311–21)

* At the time of writing, full details regarding size, dietary requirements and reproductive biology have not yet been determined. However, the extremely restricted range of this species dictates that we regard it as being highly endangered. The search for appropriate rearing and breeding methods in order to ensure the survival of this new arrival must therefore be afforded top priority

GROUP 2 (Maculatus Group)

3. *Xiphophorus andersi*
RANGE: Río Atoyac in Veracruz, Mexico
OVERALL SIZE: Males around 4.5 cm (1.8 in); females around 5.5 cm (2.2 in)

WATER REQUIREMENTS AND DIET: Generally as for *X. couchianus*
BREEDING: Average broods of around 50 fry every 5–6 weeks, but a record 111 fry have been reported by James Langhammer
NOTES: This is a rather plainly coloured fish exhibiting a delicate net-like pattern of scaling in its body and a black line extending from the anus to the caudal fin. Fertile hybrids have been produced between this species and, reportedly, *X. xiphidium*

4. *Xiphophorus evelynae*
SYNONYM: *Xiphophorus variatus evelynae*
RANGE: Río Tecolutla system in Puebla, Mexico
OVERALL SIZE: Males up to 4.5 cm (1.8 in); females up to 5.5 cm (2.2 in)
WATER REQUIREMENTS: Well-filtered, aerated water (avoiding extremes of

Plate 66   *Xiphophorus andersi* male. (*Wilf Blundell*)

Plate 67   This is a pair of fertile hybrids produced between *X. andersi* and, reportedly, *X. xiphidium*. (*Dennis Barrett*)

Plate 68   *Xiphophorus evelynae* male, collected at Necaxa in Mexico. (*Manfred Meyer*)

pH and hardness) kept between 20–24 °C (68–75 °F)

PREFERRED DIET: Live and dry food

BREEDING: Broods between 25 and 75 have been reported, but the latter figure appears to be unconfirmed and may well be excessive

NOTES: This is an attractive species which does quite well in mixed communities as long as its tankmates are not too aggressive or large. Hybridisations are likely to occur between this and other related species

5. *Xiphophorus maculatus*

SYNONYMS: *Platypoecilus maculatus, Poecilia maculata*

COMMON NAMES: Platy, Moonfish, Southern Platyfish

RANGE: From Veracruz in Mexico along the Atlantic slope down to Belize, British Honduras and Guatemala

OVERALL SIZE: Males around 3.5 cm (1.4 in); females around 6 cm (2.4 in), but often smaller

WATER REQUIREMENTS: Neutral or slightly alkaline, medium-hard water kept between 20–25 °C (68–77 °F)

PREFERRED DIET: Wide range of foods, which should include a vegetable component

BREEDING: Up to 80 fry or so produced every 4–6 weeks

NOTES: This is a very variable species. This inherent variabilty, plus the species' tendency to hybridise with other *Xiphophorus* species has been extensively exploited commercially, resulting in a wide range of colour and fin configurations, plus a whole spectrum of spectacular hybrids, mostly between it and *X. helleri* and *X. variatus*

6. *Xiphophorus milleri*

COMMON NAME: Catemaco Livebearer

RANGE: Around Laguna Catemaco in Veracruz, Mexico

OVERALL SIZE: Males up to 3.8 cm (1.5 in); females around 5 cm (2 in)

WATER REQUIREMENTS: Undemanding, but extremes of pH and hardness should be avoided. Temperatures from as low as 14.5 °C (58 °F) in winter to around 25 °C (77 °F) in summer

Plate 69   *Xiphophorus maculatus* – the Platy. This speckled specimen is from Belize. (*Manfred Meyer*)

Plate 70 *Xiphophorus maculatus* – a Red Iris male from the Belize River. (*Manfred Meyer*)

Plate 71 A splendid male *Xiphophorus maculatus*, from the Río Coatzacoalcos in Mexico. (*Dennis Barrett*)

Plate 72 A prize-winning domesticated variety of *Xiphophorus maculatus* – the Candy Platy from Segrest Farms in Florida. (*Harry Grier/Florida Tropical Fish Farms Association*)

PREFERRED DIET: Wide range of live and dry foods

BREEDING: Around 30–40 fry every 5–6 weeks

NOTES: This is a somewhat elongated species with a conspicuous small black spot on the caudal peduncle. Hybrids between it and *X. helleri* have been produced by at least one aquarist (Peter Capon) – all four survivors grew up into males

Plate 73 A Sunburst Twinbar Platy (*Xiphophorus maculatus*) produced by Gordon Aquatics in Florida. (*Harry Grier/Florida Tropical Fish Farms Association*)

Plate 74 An adult pair (male below) of the Catemaco Livebearer (*Xiphophorus milleri*). (*Dennis Barrett*)

7. *Xiphophorus variatus*
SYNONYM: *Platypoecilus variatus*
COMMON NAMES: Sunset, Variable, Variegated or Variatus Platy. Rarely: Moctezuma Platy
RANGE: Atlantic slope of eastern Mexico
OVERALL SIZE: Male about 5.5 cm (2.2 in); females up to 7 cm (2.8 in) – commercially produced varieties often larger than wild-caught specimens
WATER REQUIREMENTS: Neutral or slightly alkaline, slight- to medium-hard water. Wide temperature range – from 16 °C (61 °F) up to 27 °C (80 °F)
PREFERRED DIET: As for *X. maculatus*
BREEDING: Large broods of well over 100 fry are possible from large females but average of around 50 is far more common. Gestation period: 4–6 weeks
NOTES: This is another naturally variable species that has been widely exploited commercially. A natural hybrid between it and *X. xiphidium*, known as *X. 'kosszanderi'* is known. Aquarium hybrids between *X. variatus* and *X. maculatus* and *X. helleri*

abound, including a Florida-developed variety that is *variatus*-shaped but carries a sword

8. *Xiphophorus xiphidium*
SYNONYMS: *Xiphophorus variatus xiphidium*, *Platypoecilus maculatus*, *P. variatus*, *P. xiphidium*
COMMON NAME: Swordtail Platy
RANGE: Río Soto la Marina system in Tamaulipas, Mexico
OVERALL SIZE: Males up to 4 cm (1.6 in); females up to 5 cm (2 in)
WATER REQUIREMENTS: Neutral to slightly alkaline water kept around 25 °C (c. 77 °F)
PREFERRED DIET: As for *X. maculatus*
BREEDING: Smallish broods of around 30 or less every 5–7 weeks
NOTES: This is a slightly more difficult species to maintain than *X. maculatus* with spawnings being more unpredictable and the rearing of fry more challenging. At least two varieties exist – one with a single spot and the other with two spots on the caudal peduncle. *X. xiphidium* will hybridize with other Platy species (see, for example, *X. andersi* and *X. variatus*)

Plate 75 *Xiphophorus variatus* – the Sunset Platy. This male was collected in the Río Axtla in Mexico. (*Manfred Meyer*)

Plate 76   Some cultivated Hi-fin strains of *Xiphophorus variatus* have exceptionally long dorsal fins. (*Ross Socolof*)

Plate 77   The ease with which Platies will hybridise with Swordtails has been exploited to produce numerous varieties of fancy strains like this award-winning Rainbow Sword Variatus male from Ruskin Tropicals in Florida. (*Harry Grier/Florida Tropical Fish Farms Association*)

Plate 78 *Xiphophorus xiphidium* – the Swordtail Platy. This male belongs to the Two-spot variety. (*Dennis Barrett*)

Plate 79 A male *Xiphophorus nigrensis* from the Río Coy population in Mexico. (*Manfred Meyer*)

GROUP 3 (Helleri Group)

Subgroup A – Pygmaeus Subgroup

Members of this subgroup have blunt gonopodial blades, a relatively short sword (extremely short in *X. pygmaeus*) and a claw-like tip to ray 5a

9. *Xiphophorus nigrensis*
SYNONYM: *Xiphophorus pygmaeus nigrensis*
COMMON NAMES: Dwarf Helleri, Pygmy Swordtail*
RANGE: Ríos Choy and Coy in the Río Panuco basin of San Luis Potosí, Mexico
OVERALL SIZE: Males and females up to 6 cm (2.4 in)
WATER REQUIREMENTS: Neutral or slightly alkaline water at 24–27 °C (75–80 °F)
PREFERRED DIET: Mixed diet, incorporating livefoods
BREEDING: Not particularly easy, but small broods of around 20–25 fry produced every 6–8 weeks or so
NOTES: There is some variation in this somewhat stocky species, both in coloration and in the length of the, albeit short, sword. The sword has a pronounced black lower border

* These common names are misleading and should be discontinued or perhaps applied solely to *X. pygmaeus*

10. *Xiphophorus pygmaeus*
SYNONYM: *Xiphophorus pygmaeus pygmaeus*
COMMON NAMES: Dwarf Helleri, Pygmy Swordtail
RANGE: Río Axtla drainage system in San Luis Potosí, Mexico
OVERALL SIZE: Males around 3.5 cm (1.4 in); females up to 4.5 cm (1.8 in)
WATER REQUIREMENTS, DIET AND BREEDING: As for *X. nigrensis*
NOTES: Many *X. pygmaeus* males do not develop a sword. This is a somewhat delicate and variable species in which some males may be predominantly

yellow in base colour, while others from the same river may be bluish

Subgroup B – Montezumae Subgroup

Species belonging to this subgroup are distinguished from the preceding one in possessing double-margined swords with a tendency to curve upwards (not very detectable in *X. montezumae*)

11. *Xiphophorus birchmanni*
SYNONYM: *Xiphophorus montezumae birchmanni*
RANGE: Río Samancha in the Río Panuco basin of Huejutla in Hidalgo, Mexico
OVERALL SIZE: Males around 8 cm (3.2 in); females around 5 cm (2 in). These figures are based on the limited number of specimens raised in aquaria at the time of writing
WATER REQUIREMENTS: Alkaline, medium-hard water (pH 8.1–8.3; GH 12°; KH 12°) has been found suitable.* Temperature: 21–25 °C (70–77 °F)
PREFERRED DIET: Livefoods and vegetable-based flake have been found acceptable
BREEDING: Very small broods produced so far (around 10 fry) on an infrequent basis
NOTES: This species has only recently found its way into specialist aquarists' tanks. At the time of writing, few collections had been made, notably Hnilicka (1984), Lechner and Radda (1987), Meyer (1987) and Vernon (1988). Lechner and Radda described this fish as a subspecies of *X. montezumae* in: *Revision des* Xiphophorus montezumae/cortezi *komplexes und Neubeschreibung einer subspezies* (DGLZ – Rundschau, 2/88). Colin Vernon reports that *X. birchmanni* was never caught in open water, but rather in backwaters with little vegetation and numerous rocks. *X. variatus* and specimens that

Plate 80   A male, 'blue' morph, Pygmy Sword (*Xiphophorus pygmaeus*) from the Río Axtla in Mexico. (*Manfred Meyer*)

Plate 81   *Xiphophorus birchmanni* (male), from the Orizatlán population in Hidalgo, Mexico. (*Manfred Meyer*)

Plate 82   *Xiphophorus cortezi* (male), collected in the Río Axtla in Mexico. (*Manfred Meyer*)

appeared to be *X. birchmanni* × *variatus* hybrids were also collected

* (a) pH – This is a measure of the acidity/alkalinity of a solution. A value of pH7 is regarded as neutral. Readings below this indicate increasing levels of acidity, with values above 7 indicating increasing levels of alkalinity
(b) GH – This is a measure of the general or permanent hardness of a water sample. The readings given for *X. birchmanni* are in German degrees. To convert this to parts per million (ppm), the figure needs to be multiplied by 17.9
(c) KH – This is a measure of the bicarbonate or temporary hardness of a water sample. To obtain a figure in parts per million (ppm), the German degrees of KH are multiplied by 17.9

Total hardness of a water sample

Hardness	Parts per million (ppm)		
Very Soft	0	–	50
Moderately soft	50	–	100
Slightly hard	100	–	150
Moderately hard	150	–	200
Hard	200	–	300
Very hard	Above		300

12. *Xiphophorus cortezi*
SYNONYM: *Xiphophorus montezumae cortezi*
COMMON NAME: Montezuma Swordtail*
RANGE: Río Panuco basin; Hidalgo and San Luis Potosí, Mexico
OVERALL SIZE: Males around 4.5 cm (1.8 in), excluding sword; females up to 6 cm (2.4 in)
WATER REQUIREMENTS: Water composition not critical but good filtration and aeration recommended, along with clumps of vegetation for shelter. Temperature: 22–25 °C (72–77 °F)
PREFERRED DIET: Wide range of live and dry foods
BREEDING: Broods average around 40 fry but up to 76 have been reported (Dan Fromm). Gestation period: 5–8 weeks

NOTES: The short sword in *X. cortezi* shows a distinct tendency to curve upwards

* This common name should perhaps be discontinued and restricted to *X. montezumae*

13. *Xiphophorus montezumae*
SYNONYM: *Xiphophorus montezumae montezumae*
COMMON NAME: Montezuma Swordtail
RANGE: Río Tamesí basin in Tamaulipas and Río Panuco basin (e.g. Ríos Salto de Agua, Axtla, and Verde) in San Luis Potosí, Mexico
OVERALL SIZE: Males around 5.5 cm (2.2 in), excluding sword; females around 6.5 cm (2.6 in); these are average figures only – different populations may differ in body size and length of sword
WATER REQUIREMENTS: Water composition not too critical but soft, acid conditions should be avoided. Temperature range reported from as low as 14.5 °C (58 °F) to 25 °C (77 °F)
PREFERRED DIET: Wide range of live and dry foods
BREEDING: Broods of around 40–50 fry every 6–8 weeks
NOTES: A range of body and colour forms of this elegant species is known. There is still some controversy regarding the status of the various populations and of the species itself (regarded as a subspecies by some workers)

Subgroup C – Clemenciae Subgroup

X. clemenciae, the sole representative of this subgroup, differs from the preceding subgroup in its coloration (which resembles that of normal Swordtails). It does, however, retain the blunt gonopodial blade of Subgroups A and B.

14. *Xiphophorus clemenciae*
COMMON NAME: Yellow Swordtail
RANGE: Río Sarabia in the Río Coatzacoalcos system, Oaxaca, Mexico
OVERALL SIZE: Males around 4 cm
(1.6 in), excluding sword; females
around 5.5 cm (2.2 in)
WATER REQUIREMENTS: Well-filtered,
aerated water, avoiding low pH and
softness. Temperature: 24–26 °C
(75–79 °F)
PREFERRED DIET: Live and dry foods
accepted
BREEDING: Small broods of 20 (or fewer)
fry reported. Gestation period 5–8
weeks
NOTES: This is a particularly attractive
species of Swordtail with a rather
wide caudal peduncle, particularly in
fully mature males

Subgroup D – Alvarezi Subgroup

X. alvarezi, the sole member of Subgroup D, resembles *X. helleri* and *X.
signum* in having the terminal hook

on ray 3 shortened and strongly
curved and differs from the preceding
groups in having a pointed gonopodial blade

15. *Xiphophorus alvarezi*
SYNONYM: *Xiphophorus helleri alvarezi*
RANGE: Mainly Ríos Santo Domingo
and Palenque in Chiapas, Mexico; El
Quiché and Alta Verapaz,
Guatemala
OVERALL SIZE: Males up to 6 cm (2.4 in),
excluding sword; females 7.5 cm
(3 in)
WATER REQUIREMENTS: Neutral or alka-
line, well-filtered water at around 22–
26 °C (72–79 °F)
PREFERRED DIET: Live and dry foods,
including a vegetable-based com-
ponent
BREEDING: Broods of around 50–60 fry.
Gestation period 5–6 weeks
NOTES: This is a slender, elegant species
which is often reported as a subspecies of *X. helleri*

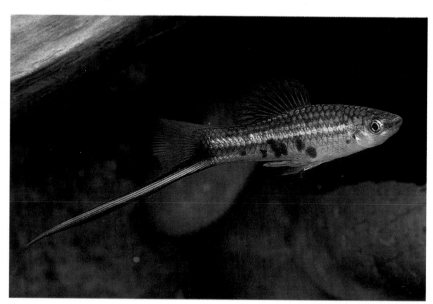

Plate 83 The true Montezuma Swordtail (*Xiphophorus montezumae*) has a long,
straight sword, beautifully illustrated in this male from Tomasopo, Mexico.
(*Manfred Meyer*)

Plate 84 The elegant Yellow Swordtail (*Xiphophorus clemenciae*). This male comes from the Río Sarabia in Oaxaca, Mexico. (*Manfred Meyer*)

Plate 85 *Xiphophorus alvarezi* – an aquarium-bred male. (*Wilf Blundell*)

Subgroup E – Helleri Subgroup

This subgroup is distinguished from the preceding ones in having a pointed gonopodial blade over which the sharply recurved tip of ray 4a fits. As in *X. alvarezi*, the claw on ray 5a is greatly enlarged

16. *Xiphophorus helleri*
SYNONYMS: *Xiphophorus helleri helleri, X. h. guentheri, X. jalapae, X. h. strigatus, X. strigatus, X. h. brevis, X. brevis, X. rachovii*
COMMON NAME: Swordtail
RANGE: Atlantic drainage from Río Nautla in Veracruz, Mexico, south to northern Honduras (see Table IX for exotic distribution details)

OVERALL SIZE: Males up to 14 cm (5.5 in), excluding sword; females up to 16 cm (6.3 in), but usually considerably smaller
WATER REQUIREMENTS AND DIET: As for *X. alvarezi*
BREEDING: Up to 150 fry or more produced every 4–6 weeks
NOTES: This is a highly variable species, some populations of which have been afforded specific or subspecific status by various workers. Body coloration can range from red to green, with or without speckling (this last trait often associated with a form previously referred to as *X. guentheri*). Numerous body and fin configurations have been developed commercially, usually involving hybridisation with *X. maculatus* and *X. variatus*

Plate 86   Numerous varieties of Swordtail (*Xiphophorus helleri*) are found in the wild. This one from Río Papaloapan is delicately coloured. (*Manfred Meyer*)

Plate 87   Belize populations of *Xiphophorus helleri* include spotted individuals like this elegant male. (*Manfred Meyer*)

Plate 88 Among the countless man-made varieties of Swordtail is this White Bloodfin Lyretail from Rainbow Fishery in Florida. (*Harry Grier/Florida Tropical Fish Farms Association*)

Plate 89 A recent arrival on the Swordtail scene – the Black Marble Sword from Florida's Gordon Aquatics. (*Harry Grier/Florida Tropical Fish Farms Association*)

Subgroup F – Signum Subgroup

The sole representative of this sub-group, *Xiphophorus signum*, exhibits the gonopodial characteristics of *X. helleri*, but differs from it in being uniformly green-gold in body colour – i.e. it lacks the typical Swordtail body stripes. *X. signum* also carries a distinctive black spot/streak in the lower rays of the caudal fin

17. *Xiphophorus signum*
SYNONYM: *Xiphophorus helleri signum*
COMMON NAME: Signum Swordtail
RANGE: Río de la Pasión basin, Alta Verapaz in Guatemala
OVERALL SIZE: Males around 7.5 cm (3 in), excluding sword; females around 10 cm (4 in)
WATER REQUIREMENTS, DIET AND BREED-ING: As for *X. helleri*, but broods are usually smaller
NOTES: This is a large, elegant fast-swimming species which will hybridise with other species of the Helleri Subgroup and, possibly, other members of the genus *Xiphophorus*

TRIBE: CNESTERODONTINI

GENUS: *Cnesterodon*: The scientific name of the genus means chisel-toothed, a reference to the scraping-knife arrangement of teeth that distinguishes these fish from all other Poeciliines

1. *Cnesterodon carnegiei*
SYNONYM: *Cnesterodon decemmacu-latus**
RANGE: Southeastern Brazil and Uruguay
OVERALL SIZE: Males around 2.5 cm (1 in); females around 3.5 cm (1.4 in)
WATER REQUIREMENTS: Chemical compo-sition not critical but extremes of pH and hardness should be avoided. Temperature: 20–24 °C (68–75 °F)

PREFERRED DIET: Live and dry foods, including a vegetable component
BREEDING: Generally small broods of around 20 fry (but 51 have been reported by James Langhammer). Gestation: 4–6 weeks
NOTES: This is one of the most southerly distributed genera of livebearers

* The synonym arose largely out of the close similarity between *C. carnegiei* and *C. decemmaculatus*. The latter, while normally bearing spots, as its name implies, can some-times appear devoid of these, depending on the mood of the fish and environmental con-ditions. There are also slight gonopodial differences between the two species. *C. decemmaculatus* has been introduced into at least one locality in Chile

GENUS: *Phalloceros*: This genus contains only one species, *P. caudimaculatus*, distinguished from other Poeciliines by the possession of a horn-like arrangement of the gonopodial tip segments

1. *Phalloceros caudimaculatus
SYNONYMS: *Phalloceros caudomacula-tus**, *Girardinus caudimaculatus*, *Poecilia caudimaculatus*, *Glari-dichthys caudimaculatus*
COMMON NAMES: One-spot Livebearer; the Caudo, Golden One-spot*; Spotted Livebearer*; Golden Spotted Livebearer*
RANGE: Southen Brazil, Uruguay and Paraguay. Introduced populations have been reported from Malawi and Western Australia
OVERALL SIZE: Males around 3.0 cm (1.2 in); females around 5 cm (2 in)
WATER REQUIREMENTS: Populations from coastal waters (usually non- or one-spotted forms) benefit from a small amount of salt in the water – about 5 ml (1 teaspoonful) per 4.5 litres (1 Imperial gallon). Other populations may be kept in non-brackish neutral to alkaline water. Temperature range from 18 °C (64 °F) to around 24 °C (75 °F)

Plate 90 *Xiphophorus signum* is characterised by a black streak in the lower rays of the caudal fin. This female comes from the Río Chacmaij population in Mexico. (*Manfred Meyer*)

Plate 91 *Cnesterodon carnegiei* – a pair (male above) from Uruguay. (*Manfred Meyer*)

PREFERRED DIET: Live and dry foods, including a vegetable component

BREEDING: Broods from 15 to 80 fry. Gestation: 5–6 weeks

NOTES: * This is a variable species, ranging in body colour and patterning from the basic non- or one-spotted wild-type forms to a one-spotted golden form, a reticulated form and a reticulated golden form. These varieties have given rise to a number of scientific names, viz. *P. c. auratus*, *P. c. reticulatus* and *P. c. r. auratus*. Most recent workers regard the above as no more than varieties of *Phalloceros caudimaculatus*

GENUS *Phalloptychus*: This genus is distinguished from other members of the tribe Cnesterodontini in having an asymmetrical gonopodium, twisted to the left

1. *Phalloptychus januarius*

SYNONYMS: *Poecilia januarius*, *Girardinus inheringii*, *Girardinus januarius*

COMMON NAMES: Barred or Striped Millions Fish

RANGE: Río de Janeiro to Río Grande do Sul in Brazil; eastern Paraguay and Uruguay

OVERALL SIZE: Males around 2.5 cm (1 in); females around 4 cm (1.6 in)

WATER REQUIREMENTS: Well-aerated water with some salt added – about 5 ml (1 teaspoonful) per 4.5 litres (1 Imperial gallon). Temperature: 20–25 °C (68–77 °F). This species is, however, present in Lago Rodrigo de Freitas in Rio de Janeiro where I have recorded a temperature of 34.4°C (94°F)

PREFERRED DIET: Live or dry foods

BREEDING: Between 10 and 30 fry produced every 4–6 weeks (fry are born over a period of a week or so rather than in a single batch)

NOTES: A second species, *P. eigenmanni*, sometimes appears in the literature, usually as a subspecies of *P. januarius*

GENUS: *Phallotorynus*: The fourth genus of this tribe, *Phallotorynus*, is from southeastern Brazil and Paraguay. Neither of its two constituent species *P. fasciolatus* or *P. jucundus*, are known to any degree within the aquarium hobby

TRIBE: SCOLICHTHYINI

GENUS: *Scolichthys*: While appearing superficially like *Cnesterodon*, the genus *Scolichthys* may be distinguished by its less elaborate bony 'style' of more or less rigid membranous tissue located at the tip of the gonopodium. *Scolichthys* also lacks the well-formed claw that *Cnesterodon* possesses on gonopodial ray 5a

1. *Scolichthys greenwayi*

RANGE: Río Usumacinta basin in Alta Verapaz, Guatemala

OVERALL SIZE: Males up to 5 cm (2 in); females around 5 cm (2 in)

WATER REQUIREMENTS: Neutral to alkaline, medium-hard water. Temperature range: 22–30 °C (72–86 °F) – but around 25 °C (c. 77 °F) is preferable

PREFERRED DIET: Mostly livefoods

BREEDING: Average broods of 10–30 fry produced at 5–7 week intervals, but 75 reported by aquarist Mike Schadle

NOTES: This is an attractive fish distinguished by a central black body spot present in both sexes. The genus is the only new one erected since Rosen and Bailey's major revision of 1963. A second species, *S. iota*, does not, as yet, appear to have found its way into the aquatic hobby

TRIBE: GAMBUSINI

GENUS: *Belonesox*: The most distinctive feature of this fish is a prominent beak-like mouth with rows of sharp pointed teeth

1. *Belonesox belizanus*

COMMON NAMES: Pike Killifish, Pike Top Livebearer, Pike Topminnow

Plate 92 *Phalloceros caudimaculatus* – a one-spotted male from Brazil. (*Manfred Meyer*)

Plate 93 *Phalloceros caudimaculatus 'reticulatus auratus'*. (*Dennis Barrett*)

Plate 94   *Scolichthys greenwayi* – female showing the characteristic elongated body spot. (*Derek Lambert*)

Plate 95   *Belonesox belizanus* (male) – the Pike Top Livebearer. (*Bill Tomey*)

Plate 96   *Brachyraphis episcopi* – the Bishop. This male was collected in Panama. (*Manfred Meyer*)

RANGE: From Veracruz (Mexico) southwards to Yucatán, Guatemala, Honduras and Nicaragua. Introduced into parts of Florida and probably San Antonio River (Texas). See Table IX for further details of exotic distribution

OVERALL SIZE: Males up to around 12 cm (4.7 in); females much larger – up to 20 cm (8 in)

WATER REQUIREMENTS: Well-filtered water; neutral to alkaline conditions and slightly salty, hard water. Temperature around 26 °C (79 °F); slightly higher for breeding purposes. Well-planted aquaria recommended

PREFERRED DIET: Exclusively animal-based. Will take livefoods, freeze-dried and deep frozen foods

BREEDING: Large – around 2.5 cm (1 in) – fry produced in batches of up to around 100 every 4½–7 weeks. Some reports state that females will not eat their fry but these should be treated with caution

GENUS: *Brachyraphis*: This genus of stocky-looking fish shares the cannibalistic and aggressive tendencies of its near relative, *Gambusia*. However, *Brachyraphis* species are generally more colourful and possess a net-like scale pattern

1. *Brachyraphis episcopi*
SYNONYMS: *Gambusia episcopi*, *Gambusia latipunctata*, *Priapichthys episcopi*
COMMON NAME: The Bishop
RANGE: Atlantic and Pacific slopes of central Panama
OVERALL SIZE: Males 3–3.5 cm (1.2–1.4 in); females 4.5–5 cm (1.8–2 in)
WATER REQUIREMENTS: Not critical. Temperature around 24 °C (75 °F) for maintenance and slightly higher for breeding
PREFERRED DIET: Most foods
BREEDING: Relatively small broods – around 20 fry every 4 weeks or so
NOTES: Parents are highly cannibalistic and need to be separated from their

fry. Like other *Brachyraphis* fry, *B. episcopi* young carry a distinctive black spot on the vent, making identification easy even in a mixed batch.
B. episcopi is not a good community species, being somewhat aggressive and exhibiting fin-nipping tendencies

2. *Brachyraphis hartwegi*
RANGE: Pacific drainage of Chiapas, Mexico and Guatemala
OVERALL SIZE: Males about 3.5 cm (1.4 in); females about 5 cm (2 in)
WATER REQUIREMENTS, DIET AND BREEDING: Basically similar to *B. episcopi*
NOTES: *B. hartwegi* is a somewhat slender species which can be variable in coloration and is sometimes less brightly coloured than some of the other *Brachyraphis* species

3. *Brachyraphis rhabdophora*
SYNONYMS: *Gambusia rhabdophora*, *Priapichthys olomina*, *Panamichthys tristani* and others
RANGE: Atlantic and Pacific slopes of Costa Rica
OVERALL SIZE: Males up to 4 cm (1.6 in); females up to 7 cm (2.75 in)
WATER REQUIREMENTS, DIET AND BREEDING: Basically as for *B. episcopi*
NOTES: Brightly coloured specimens of this species are about the most attractive in the genus. Fry, like all other *Brachyraphis* young, are slow-growing

4. *Brachyraphis roseni*
COMMON NAME: Cardinal Brachy (name coined in 1987 by the British aquarist Ivan Dibble)
RANGE: Southeastern Costa Rica, western Panama
OVERALL SIZE: Males around 5 cm (2 in); females 6.5 cm (2.5 in)
WATER REQUIREMENTS: Well-filtered water and partial changes beneficial. Temperature between 22–30 °C (71.5–86 °F)
DIET AND BREEDING: As for *B. episcopi*
NOTES: This is a very colourful, recently described (Bussing, 1988) species. It

has been reported that fry tend to be produced either during the night or very early in the morning

5. OTHER *Brachyraphis* SPECIES

The following species of *Brachyraphis* are currently recognised, although, with the possible exception of *B. terrabensis*, none are kept in any large numbers:

> *Brachyraphis cascajalensis* Atlantic slope of southeast Costa Rica, eastwards to Río Nargana in Panama; sometimes found in brackish water
> *Brachyraphis holdridgei* Río Madre de Dios, Río Sarapiqui and Río Arenal on the Atlantic slope of Costa Rica
> *Brachyraphis parisima* Atlantic lowland coastal areas of Costa Rica
> *Brachyraphis punctifer* Atlantic slope of western Panama
> *Brachyraphis terrabensis* Pacific slope of southern Costa Rica, western Panama

GENUS: *Gambusia*: Males of the largely predatory species that make up this genus all share the characteristic that the first 4–6 rays of the pectoral fin are thickened and curved upwards on their distal half (the half furthest from the body) to form a bow or notch, used to aid stabilisation of the gonopodium in its anterior swing during copulation

A. Subgenus: *Arthrophallus*

1. *Gambusia affinis*

SYNONYMS: *Gambusia affinis affinis*, *G. speciosa*, *G. gracilis*, *Heterandria affinis*, *Zygonectes atrilatus* and others

COMMON NAME: Western Mosquito Fish

RANGE: Río Panuco basin, northern Veracruz in Mexico; northwards to southern Indiana and east to Alabama, including Mississippi drainage system and Texas. Numerous exotic populations established elsewhere (see Table IX)

OVERALL SIZE: Males 3–4 cm (1.2–1.6 in); females 5–7 cm (2–2.8 in)

WATER REQUIREMENTS: Wide range of chemical conditions tolerated. Temperature tolerance from near-freezing to over 30 °C (86 °F)

PREFERRED DIET: Largely carnivorous, but will accept virtually any other food offered

BREEDING: Wide range in size of broods from as little as 10 fry to around 80, produced every 5–8 weeks

NOTES: This species has traditionally been known as *G. affinis affinis*, i.e. it has been regarded as a subspecies – its 'sister' subspecies being *G. a. holbrooki*. Wooten, Scribner and Smith (*Copeia*, 1988(2), pp 283–9) put forward a strong argument – accepted here – for its elevation to specific level. Conclusions based on genetic variability and reproductive barriers (hybrids between *affinis* and *holbrooki* tend to be deformed or die an early death) add considerable strength to the argument. All populations west of Mobile Bay in Alabama should now be regarded as *G. affinis* according to Wooten and his co-workers. All populations east of this locality should be regarded as *G. holbrooki*. Taken together, both species almost certainly have the widest distribution of any fish in the world

2. *Gambusia atrora*

RANGE: Río Axtla and Río Matlapa in San Luis Potosí, Mexico

OVERALL SIZE: Males 3 cm (1.2 in); females 4 cm (1.6 in)

WATER REQUIREMENTS: Not too particular. Temperature around 25 °C (77 °F)

PREFERRED DIET: As for *G. affinis*

BREEDING: Smaller broods than *G. affinis*

NOTES: A species with a somewhat restricted range; basically unspectacular but with attractive reflective scales and black borders to the dorsal and caudal fins

Plates 97a and b *Brachyraphis hartwegi* (male, top, and female, below). (*Dennis Barrett*)

Plate 98   *Brachyraphis rhabdophora* (male), from Costa Rica. (*Manfred Meyer*)

Plate 99   *Brachyraphis roseni* – the Cardinal Brachy. A beautiful male from Golfito in Costa Rica. (*Manfred Meyer*)

Plate 100   The Western Mosquito Fish, *Gambusia affinis* (male) – the most significant fish (in conjunction with *G. holbrooki*) used in the biological control of malaria, with dubious benefits. (*John Dawes*)

Plate 101 *Gambusia atrora* pair (male below), from Río Axtla in Mexico. (*Manfred Meyer*)

3. *Gambusia aurata*
SYNONYM: *Gambusia myersi*
COMMON NAME: Golden Gambusia
RANGE: Río Tamesí basin, Mexico
OVERALL SIZE: Males 2.5 cm (1 in); females 4 cm (1.6 in)
WATER REQUIREMENTS, DIET AND BREEDING: Generally as for *G. atrora*
NOTES: This is a very attractive *Gambusia* which, unfortunately, exhibits the fin-nipping and cannibalistic characteristics of the genus

4. *Gambusia holbrooki*
SYNONYMS: *Gambusia affinis holbrooki*, *Heterandria holbrooki* and others
COMMON NAME: Eastern Mosquito Fish
RANGE: From Central Alabama east to Florida and northwards along the Atlantic coastal drainages up to New Jersey. Numerous exotic populations established (see Table IX)
OVERALL SIZE: Males 3–4 cm (1.2–1.6 in); females up to around 7 cm (2.8 in)
WATER REQUIREMENTS, DIET AND BREEDING: As for *G. affinis*
NOTES: *G. holbrooki* males have a tendency towards melanism (either black spots or totally black). Ray 3 of the gonopodium carries denticles while the corresponding ray in *G. affinis* is smooth. Ray 4p has a short, unsegmented claw in *G. holbrooki* and a long, segmented one in *G. affinis*. See entry for *G. affinis* for further details

Plate 102   The Golden Gambusia, *Gambusia aurata*. This male comes from the Río Maute in Mexico. (*Manfred Meyer*)

Plate 103   The Eastern Mosquito Fish, *Gambusia holbrooki*. Many populations, including the one from which this pair was collected, have mottled or totally black males. (*Manfred Meyer*)

5. *Gambusia gaigei**

SYNONYM: *Gambusia alvarezi**, *Gambusia hurtadoi**

COMMON NAME: Big Bend Gambusia

RANGE: a) *G. alvarezi* – original description based on fish collected at El Ojo de San Gregorio in Chihuahua, Mexico

b) *G. gaigei* – collected at Rio Grande in Boquillas, Brewster County, Texas

c) *G. hurtadoi* – original description based on fish collected at El Ojo de la Hacienda Dolores in Chihuahua, Mexico

OVERALL SIZE: Males around 2.5–3 cm (1–1.2 in); females around 4 cm (1.6 in)

WATER REQUIREMENTS, DIET AND BREEDING: Generally as for other *Gambusia* species

NOTES: * The three names used here are regarded by some workers (including Mary Rauchenberger and Manfred Meyer) as representing three separate species, while others see them as slightly different forms of the same species. At least two populations, *G. alvarezi* and *G. gaigei*, are regarded as being endangered in the wild, although the existence of refuge populations is playing a crucial part in their continued survival

6. *Gambusia longispinis*

RANGE: Cuatro Ciénegas basin, Coahuila in Mexico

OVERALL SIZE: Males around 3 cm (1.2 in); females around 5 cm (2 in)

WATER REQUIREMENTS, DIET AND BREEDING: As for *G. atrora*

NOTES: *G. longispinis* has sometimes been confused with *G. marshi* – something that has not helped in the conservation of the former, which is regarded as threatened in the wild. Fortunately, pure stocks of *G. longispinis* now appear to be available

Plate 104 *Gambusia alvarezi* (female), from San Gregorio in Chihuahua, Mexico. (*Manfred Meyer*)

Plate 105   *Gambusia longispinis* (female), from the Cuatro Ciénegas basin in Mexico. (*Manfred Meyer*)

B.   Subgenus: *Heterophallina*

7.   *Gambusia marshi*
RANGE: Cuatro Ciénegas basin, Coahuila in Mexico
OVERALL SIZE: Males 3–3.5 cm (1.2–1.4 in); females around 6 cm (*c.* 2.4 in)
WATER REQUIREMENTS, DIET AND BREEDING: As for *G. atrora*
NOTES: Although sometimes confused with *G. longispinis*, *G. marshi* is usually larger and more robust. It is also more variable than *G. longispinis* and exhibits distinct gonopodial differences

8.   *Gambusia rachowi**
SYNONYMS: *Heterophallus rachowi**, *Gambusia atzi*
RANGE: Arroyos (brooks) Santiago Vasques and Jesús Carranza, plus Laguna de la Sapote in Veracruz, Mexico
OVERALL SIZE: Males about 2.5 cm (1 in) or slightly larger; females around 3.5 cm (1.4 in)
WATER REQUIREMENTS: Well-lit, good-quality water seems to be preferred. Temperature 22–25 °C (72–77 °F) for maintenance; slightly higher for breeding
PREFERRED DIET: Omnivorous (live/flake/vegetable foods)
BREEDING: Small broods of around 20 fry produced every 4–6 weeks
NOTES: This is a variable but attractive fish

* Some workers regard *Heterophallus* as being the correct generic name for this fish, e.g. Meyer, Wischnath and Foerster, 1985: *Lebendgebärende Zierfische* (Published by Mergus). Irrespective of the validity of this, there is no denying the *Gambusia* affinities of the species. The name used above is the one adopted by Mary Rauchenberger in 1989: 'Annotated Species List of the Subfamily Poeciliinae' in *Ecology and Evolution of Livebearing Fishes* (Published by Simon & Schuster)

Plate 106   *Gambusia marshi* (male), from Cuatro Ciénegas in Chihuahua, Mexico. (*Manfred Meyer*)

Plate 107   *Gambusia rachowi* (male), from the Jesús Carranza population in Mexico. (*Manfred Meyer*)

Plate 108   *Gambusia regani* (male), from Río Forlon, Mexico. (*Manfred Meyer*)

Plate 109   *Gambusia vittata* (male), from the Río Axtla population in Mexico. (*Manfred Meyer*)

Plate 110   *Carlhubbsia stuarti* (male) – aquarium specimen. (*Dennis Barrett*)

9. *Gambusia regani*

RANGE: Río Tamesí system on the Atlantic slope of Mexico

OVERALL SIZE: Males 2–3.5 cm (0.8–1.4 in); females 3.5–4.5 cm (1.4–1.8 in)

WATER REQUIREMENTS, DIET AND BREEDING: Generally as for *G. atrora*

NOTES: A yellowish species with a line running the length of the body. Males can differentiate at a very small size

10. *Gambusia vittata**

SYNONYMS: *Flexipenis vittatus**, *Flexipenis vittata*

RANGE: Mexico: Río Panuco basin from Ciudad Victoria in Tamaulipas to northern Veracruz

OVERALL SIZE: Males about 4.5 cm (1.8 in); females about 6 cm (2.4 in)

WATER REQUIREMENTS, DIET AND BREEDING: Similar to *G. rachowi*

NOTES: This is a gorgeous fish when in peak condition and seen in the right light

* Some workers, among them Meyer *et al.* (see *G. rachowi*), regard this fish as *Flexipenis vittatus*. The name used above is the one adopted by Mary Rauchenberger (see *G. rachowi* for full reference)

11. OTHER *Gambusia* SPECIES

There are quite a few other *Gambusia* species kept by specialist aquarists in various countries. Among them are (depending on possible synonymies):

Subgenus *Heterophallina*: *G. panuco*, *G. (Heterophallus) echeagayani*, *G. (H.) milleri*

Subgenus *Arthrophallus**: *G. speciosa*, *G. lemaitrei*, *G. nobilis***, *G. georgei***, *G. heterochir***, *G. krumholzi*, *G. sexradiata*, *G. eurystoma*, *G. senilis***, *G. geiseri***

Subgenus *Gambusia*: *G. nicaraguensis*, *G. wrayi*, *G. melapleura*, *G. puncticulata****, *G. yucatana*, *G. hispaniolae*, *G. dominicensis*, *G. hubbsi****, *G. manni****, *G. monticola****, *G. baracoana****, *G. bucheri****, *G. oligosticta****,

*G. caymanensis****, *G. howelli****, *G. luma*, *G. beebei*, *G. pseudopunctata*, *G. punctata*, *G. rhizophorae*, *G. xanthosoma*

* A member of this subgenus, *G. amistadensis* was reported extinct in 1984 – Hubbs and Jensen (*Copeia* No. 2, 1984) – through contamination of the few remaining populations with *G. affinis* and *G. gaigei*

** These species, and perhaps others, are under some form of threat in the wild

*** According to Fink 1971 – 'A Revision of the *Gambusia nicaraguensis* Species Group (Pisces: Poeciliidae)', Publ. Gulf Coast Res. Lab. Mus. 2: 47–77 – all these species were synonymised as *G. puncticulata*. Other researchers, among them Mary Rauchenberger, feel that the move was premature

TRIBE: GIRARDININI

GENUS: *Carlhubbsia*: Males of this genus have dextrally (right-handedly) asymmetrical gonopodia. They also possess serrae (saw-like structures) on ray 5p of the gonopodium, a characteristic they share with the other genera of the tribe. This feature is absent in all other Poecilliines

1. *Carlhubbsia stuarti*

RANGE: Río Polochic system and Laguna Izabal, Guatemala

OVERALL SIZE: Males up to 5 cm (2 in); females up to 6 cm (2.4 in)

WATER REQUIREMENTS: Slightly alkaline and hard water seems to suit this species best. Temperature: 24–28 °C (75–82 °F)

PREFERRED DIET: Live and dry foods

BREEDING: As few as 10 or as many as 50 fry produced every 5–7 weeks or so

NOTES: This very attractive species is sometimes confused with *Phallichthys amates* but can easily be distinguished from it by its distinct vertical stripes. Under the microscope, gonopodial ray 5p of *Phallichthys* can be seen not to carry the serrae which are characteristic of *Carlhubbsia*. A second species of *Carlhubbsia*, *C. kidderi*, from Campeche in Mexico and El Petén and Alta Verapaz in Guatemala, is also available within the aquarium hobby. *C. kidderi* is generally less colourful than *C. stuarti*

GENUS: *Girardinus*: While males of this genus possess the tribal characteristic of serrae on gonopodial ray 5p, they can be distinguished from *Carlhubbsia* in having symmetrical (as opposed to asymmetrical) gonopodia. They also possess two antrorse (directed forwards or upwards) 'membranous appendages arising ventrally at (the) level of (the) spinules' on ray 3 of the gonopodium (Rosen and Bailey, 1963)

1.   *Girardinus creolus*
SYNONYMS: *Toxus creolus*, *Toxus riddlei*
COMMON NAME: Creole Topminnow
RANGE: Sierra de los Organos, Pinar del Río in Cuba
OVERALL SIZE: Males around 4 cm (1.6 in); females around 7 cm (2.8 in)
WATER REQUIREMENTS: Well-filtered water is desirable. Temperature: 21–28 °C (70–82 °F)
PREFERRED DIET: Live and dry foods, including a vegetable component (diet based on the requirements of *G. metallicus*)
BREEDING: No details are available at the time of writing, but probably similar to *G. metallicus* with, perhaps, smaller broods
NOTES: This is probably the most attractive of the *Girardinus* species. Unfortunately, it is not yet well known within the aquarium hobby

2.   *Girardinus metallicus*
SYNONYMS: *Girardinus garmani*, *G. pygmaeus*

COMMON NAMES: The Girardinus, Metallic Topminnow
RANGE: Variety of waters in most (except extreme east) of Cuba
OVERALL SIZE: Males up to 5 cm (2 in); females up to 9 cm (3.5 in)
WATER REQUIREMENTS: Well-filtered water is recommended, with regular partial water changes. Temperature: 22–25 °C (72–77 °F), slightly higher for breeding
PREFERRED DIET: Live and dry foods, including a vegetable component
BREEDING: While over 100 fry have been reported, average broods are around the 50 mark. Gestation: 5–8 weeks (or even longer at low temperatures)
NOTES: This is the most popular of the *Girardinus* species. Cannibalism of fry may occur, so remedial steps are recommended at such times

3.   OTHER *Girardinus* SPECIES
Other species of this genus, along with their main distribution details and common names (derived from 'List of Accepted Common Names of Poeciliid Fishes' by Gary K. Meffe in *Ecology and Evolution of Livebearing Fishes*, Simon & Schuster, p 450, 1989) are:

Girardinus cubensis (Cuban Topminnow) Los Palacios in Pinar del Río, Cuba
Girardinus denticulatus (Toothy Topminnow) Remedios in Oriente, Cuba
Girardinus falcatus (Goldbelly Topminnow) San Cristobal in Pinar del Río, Cuba
Girardinus microdactylus (Smallfinger Topminnow) Sierra de los Organos in Pinar del Río, Cuba, and Isles of Pines
Girardinus serripenis (Serrated Topminnow) Río Taco Taco in Pinar del Río, Cuba
Girardinus uninotatus (Single-spot Topminnow) Sierra de los Organos in Pinar del Río, Cuba

Plate 111   *Girardinus creolus* (male), from Río Soron in Cuba. (*Manfred Meyer*)

Plate 112   *Girardinus metallicus* pair (male below). (*Derek Lambert*)

Of these, only *G. falcatus* is seen with any regularity within the aquarium hobby

GENUS: *Quintana*: Males of this monotypic (single-species) genus are distinguished from other Girardinini in possessing bilaterally symmetrical gonopodial rays, 3, 4a and 5, while ray 4p is sinistrally (left handedly) asymmetrical

1.   *Quintana atrizona*
COMMON NAMES: Black-barred Livebearer, Barred Topminnow
RANGE: Western Cuba
OVERALL SIZE: Males up to 2.5 cm (1 in); females 4 cm (1.6 in)
WATER REQUIREMENTS: Densely planted aquarium. Clean water. Temperture: 24–28 °C (75–82 °F)
PREFERRED DIET: Small live and dry foods, plus a vegetable component
BREEDING: Over 50 fry have been reported from a single female by aquarists, e.g. Charles Bialon and C. L. Michaels. Average broods of 25–30 are, however, more common. Gestation period: 5–8 weeks
NOTES: This is a timid species that is easily scared. A densely planted, gently aerated species aquarium – or one housing peaceful, small tankmates – is therefore recommended

TRIBE: HETERANDRINI

GENUS: *Heterandria*: When species as apparently different as *Heterandria formosa* (the Mosquito Fish) and *H. bimaculata* (the Two-spotted Livebearer) are assigned to the same genus, one is tempted to question the validity of this decision. According to Rosen ('Fishes from the Uplands and Intermontane Basins of Guatemala: Revisionary Studies and Comparative Geography', *Bulletin of the American Museum of Natural History*, Vol. 162, Article 5, 1979),

the following characters, among others, are sufficiently significant to warrant the inclusion of both fish in the same genus: '. . . a downturned membranous sheath at the tip of the gonopodium that incorporates the distal part of ray 4a; gonopodial ray falling short of tip of fin and bearing blunt, peglike, ventral spines distally; a series of from 7 to 15 serrae (sawlike structures) subdistally on ray 4p; ray 6 of gonopodium swollen and ankylosed (joined/fused) distally and with strong distal spur directed obliquely toward base of ray 5. . . .'

Having said all this, the facts that *H. formosa* is the only representative of the genus found in North America, and that it is the only species that exhibits superfoetation (fry produced in small numbers over an extended period of time) sets it apart from all the other species. In addition, all the others look more similar to each other than they do to *H. formosa*.

As a result, the genus is currently subdivided into two subgenera, viz *Heterandria* and *Pseudoxiphophorus*. Whether these will eventually be elevated to full generic status or not, only time will tell

Subgenus: *Heterandria*

1.   *Heterandria formosa*
SYNONYMS: *Girardinus formosus*, *Gambusia formosa*
COMMON NAMES: Mosquito Fish, Dwaf Livebearer, Dwarf Topminnow
RANGE: Southeastern North Carolina southwards through eastern and southern Georgia, Florida, Gulf Coast (to New Orleans) and Louisiana
OVERALL SIZE: Males up to 2 cm (0.8 in) – this makes them one of the smallest vertebrates known to science; females up to 3 cm (1.2 in)
WATER REQUIREMENTS: Tolerant of wide range of conditions, but alkaline, hardish conditions with some tannin

Plate 113   *Quintana atrizona* pair (male below). (*Derek Lambert*)

in the water suits this species very well. Temperature: from as low as 16 °C (61 °F) to around 25 °C (77 °F)

PREFERRED DIET: Small live and dry foods, including a vegetable component

BREEDING: Birth extends over a period of 10–14 days during which about 20 fry are produced

NOTES: Small, heavily planted species tanks are recommended

Subgenus: *Pseudoxiphophorus*

1. *Heterandria bimaculata*

SYNONYMS: *Xiphophorus bimaculatus, Poeciloides bimaculatus, Pseudoxiphophorus bimaculatus, Gambusia bimaculata* and others

COMMON NAMES: Two-spot Livebearer, Pseudo Helleri

RANGE:: Southern Veracruz and Oaxaca, Mexico, southwards along the northwestern margin of the Yucatán Peninsula. Also found in Alta Verapaz in Guatemala, Belize and Honduras

OVERALL SIZE: Males around 7 cm (2.8 in); females around 15 cm (6 in)

WATER REQUIREMENTS: Densely planted large aquarium. Well-filtered, aerated water between 22–26 °C (72–79 °F)

PREFERRED DIET: Livefoods

BREEDING: As many as 112 fry have been reported (Mike Schadle). Gestation period: 4–6 weeks. Fry are produced in broods rather than in the conveyor-belt manner of *H. formosa*

NOTES: This is a very aggressive species which should be kept either on its own or with robust similarly sized fish

2. OTHER *Heterandria (Pseudoxiphophorus)* SPECIES

Of all the other species in this subgenus, only *H. (Ps.) jonesi* is seen with any regularity in the aquarium hobby. Apart from gonopodial differences (e.g. *H. (Ps.) bimaculata* has a distinct recurved hook at the tip of gonopodial ray 4a, while *H. (Ps.) jonesi* does not), the single most obvious difference between them is the longer-based dorsal fin of *H. (Ps.) bimaculata*. All the *Heterandria (Pseudoxiphophorus)* species show intraspecific variation – differences in body form and pigmentation within each species.

Heterandria (Pseudoxiphophorus) anzuetoi Río Montagua basin in Guatemala, Copán in Honduras, Río Lempa basin in Guatemala and possibly El Salvador

Heterandria (Pseudoxiphophorus) attenuata Río Candelaria Yalicar in Alta Verapaz, Guatemala

Heterandria (Pseudoxiphophorus) cataractae Arroyo Salsicha in Alta Verapaz, Guatemala

Heterandria (Pseudoxiphophorus) dirempta Río Chajmaic in Alta Verapaz, Guatemala

Heterandria (Pseudoxiphophorus) jonesi Río Guayalejo in Río Tamesí basin southwards to Ríos Nautla and Atoyac in Veracruz, Mexico

Heterandria (Pseudoxiphophorus) litoperas Río Polochic basin in Alta Verapaz and Izabal, Guatemala

*Heterandria (Pseudoxiphophorus) obliqua Río Dolores system (with a subterranean link to Río Salinas) in Alta Verapaz, Guatemala, and Río San Ramón system (with subterranean link to Ríos Ixcan and Lacantún) in Huehuetenango, Guatemala

* Specimens appearing to be hybrids between *H. (Ps.) obliqua* and *H. (Ps.) bimaculata* have been collected in the upper parts of the Río Lacantún and Río Salinas basins, in Huehuetenango and in El Quiché

GENUS: *Neoheterandria*: The most easily detectable feature of the gonopodium of *Neoheterandria* males is a long, consolidated bony rod that projects forward and downward at the tip of ray 4a

1. *Neoheterandria tridentiger*

SYNONYMS: *Gambusia tridentiger, Priapichthys tridentiger, Allogambusia tridentiger*

RANGE: Panama – Canal Zone, Fort Sherman, Ríos Chame and Arrijan and south of Toro Point

OVERALL SIZE: Males up to 3 cm (1.2 in); females up to 5 cm (2 in)

WATER REQUIREMENTS: This species has been collected in water with a pH value of 4.3, but it will live quite happily in neutral or alkaline water. Temperature: 25–33 °C (77–91 °F), but the top end of the range should be avoided

PREFERRED DIET: Live and dry foods

BREEDING: Young females produce fry at the rate of about two or three per day over a period of about one week (this is a good example of superfoetation). However, as they grow older, they tend to produce fry in complete broods of around 25 fry every month or so

NOTES: Superficially, this fish could be confused with *Gambusia affinis*. However, it can be easily distinguished by its faint vertical body stripes

2. OTHER *Neoheterandria* SPECIES

Two other species, *N. cana* (from Panama) and *N. elegans* (from Colombia) are also available from time to time, the latter currently becoming established within specialist circles in the US, West Germany and the UK

GENUS: *Phallichthys*: Males of this genus have gonopodia that are sinistrally or dextrally asymmetrical. Superficially similar to *Carlhubbsia*, but *Phallichthys* species do not possess serrae (saw-like structures) on ray 5p of the gonopodium

1. *Phallichthys amates*

This species is currently accepted as consisting of two subspecies – *P. amates amates* and *P. amates pittieri*.

Plate 114 *Heterandria formosa* female. This species is commonly known as the Mosquito Fish, a name it shares with *Gambusia affinis* and *G. holbrooki*. (*Dennis Barrett*)

Plate 115 *Heterandria (Pseudoxiphophorus)* bimaculata (male), from Lake Catemaco in Mexico. (*Manfred Meyer*)

Plate 116 *Neoheterandria tridentiger* (female) – aquarium specimen. (*Dennis Barrett*)

But, should Rosen's 1979 arguments in his 'Fishes from the Uplands and Intermontane Basins of Guatemala: Revisionary Studies and Comparative Geography' (*Bulletin of the American Museum of Natural History*, Vol. 162, Article 5) regarding the validity of the subspecies concept be accepted (and there seem to be some strong reasons for doing so), then the above fish should be regarded as separate full species in their own right. However, pending that decision, I will deal with them here in the traditional manner

(i) Subspecies: *Phallichthys amates amates*
SYNONYMS: *Poecilia amates, Poeciliopsis amates*
COMMON NAME: Merry Widow
RANGE: Atlantic slope of southern Guatemala and northern Honduras
OVERALL SIZE: Males up to 4 cm (1.6 in); females up to 7 cm (2.8 in)
WATER REQUIREMENTS: Quiet conditions; vegetated aquarium. Water conditions not too critical, if extremes of pH and hardness are avoided. Temperature: 22–28 °C (72–82 °F)
PREFERRED DIET: Live and dry foods, including a vegetable component
BREEDING: As many as 151 fry have been reported (Ivan Dibble), but 50 is a more common figure. Gestation period: 4–6 weeks
NOTES: This is a quiet fish. Water turbulence and boisterous tankmates should therefore be avoided. Under appropriate conditions, *P. a. amates* is extremely attractive

(ii) Subspecies: *Phallichthys amates pittieri*
SYNONYMS: *Poecilia pittieri, Poeciliopsis pittieri, Phallichthys pittieri, Poeciliopsis isthmensis, Phallichthys ismenthis*
COMMON NAME: Orange-dorsal Livebearer
RANGE: Atlantic slope of Costa Rica and western Panama; Río Huahuasán in Zelaya, Nicaragua

OVERALL SIZE: Males up to 6 cm (2.4 in); females reported up to 10 cm (4 in)
WATER REQUIREMENTS, DIET AND BREEDING: Generally as for *P. a. amates*, though broods can be larger
NOTES: This peaceful fish grows to a larger size than its sister subspecies and is generally bluer in body colour with a somewhat less distinct black/yellow edge to the dorsal fin

2. *Phallichthys quadripunctatus*
RANGE: Río Sixaola basin, Costa Rica
OVERALL SIZE: Males around 1.5 cm (0.6 in); females around 3.5 cm (1.4 in)
WATER REQUIREMENTS: Chemical composition not critical. Wide temperature range from around 20 °C (68 °F) to over 34 °C (93 °F) reported. Around 25 °C (*c.* 77 °F) is a happy medium
PREFERRED DIET: Small live and dry foods
BREEDING: Smallish broods, but 29 fry reported by Derek Lambert. Gestation period, dependent on temperature, around 5 weeks
NOTES: This is a very small, peaceful species which should be kept either in a species tank or in one housing other small, non-boisterous fish

3. OTHER *Phallichthys* SPECIES
Two other species are recognised in this genus. *P. fairweatheri* (from the Río Hondo and New River systems of northern British Honduras, and the Río de la Pasión and Río San Pedro de Mártir systems in northern Guatemala) is an *amates*-type fish that is fairly regularly available. *P. tico* is hardly ever seen in the aquarium hobby

GENUS: *Poeciliopsis*: Males of this genus have extremely long, pointed gonopodia which exhibit sinistral asymmetry. The genus is subdivided by some workers into two subgenera, *Aulophallus* and *Poeciliopsis*, largely on the basis of gonopodial differences

Plate 117   The Merry Widow – *Phallichthys amates amates* (male). (*Dennis Barrett*)

Plate 118   *Phallichthys amates pittieri* (male) – aquarium specimen. (*Dennis Barrett*)

Plate 119   *Phallichthys quadripunctatus* (male). (*Wilf Blundell*)

Plate 120   *Poeciliopsis fasciata* (male), collected at Salina Cruz in Mexico. (*Manfred Meyer*)

Subgenus: *Poeciliopsis*
1. *Poeciliopsis fasciata*
SYNONYMS: *Gambusia fasciata, Heterandria fasciata*
RANGE: Fresh and brackish waters along the Pacific slope of Mexico (Laguna Coyuca in Guerrero to Río Pijijiapan in Chiapas); also found in the drainage headwaters of Río Coatzacoalcos, Oaxaca, Mexico
OVERALL SIZE: Males around 3 cm (1.2 in); females around 5 cm (2 in)
WATER REQUIREMENTS: Fresh or brackish water – 5 ml (1 teaspoonful) of salt per 4.5 litres (1 Imperial gallon) – neutral to alkaline, soft to slightly hard composition. Temperature: 24–26 °C (75–79 °F)
PREFERRED DIET: Live and dry foods
BREEDING: Between 15 and 30 fry every 5 weeks or so
NOTES: Until the mid-1980s, only females of this species had been collected

2. *Poeciliopsis gracilis*
SYNONYMS: *Gambusia heckeli, Xiphophorus gracilis, Girardinus pleurospilus, Heterandria pleurospilus, Poeciliopsis pleurospilus, Poecilistes pleurospilus* and others
COMMON NAME: Porthole Livebearer
RANGE: Southern Mexico to Honduras, on both the Atlantic and Pacific slopes
OVERALL SIZE: Males up to 4 cm (1.6 in); females up to 7 cm (2.8 in)
WATER REQUIREMENTS: Densely planted aquaria (but with available swimming space). Neutral to slightly alkaline water. Temperature: 22–28 °C (72–82 °F), the higher end of the scale being better suited for breeding purposes
PREFERRED DIET: Live and dry foods, including a vegetable component
BREEDING: Typical superfoetation, with a few fry produced every day over about 10 days. Maximum number reported: 60 fry (Mike Schadle)

NOTES: This is a variable species in terms of number of body spots, which can range from 5 to 10

3. *Poeciliopsis occidentalis*
SYNONYMS: *Heterandria occidentalis, Poecilia occidentalis, Mollienisia occidentalis, Girardinus occidentalis, G. sonoriensis*
COMMON NAME: Gila Topminnow
RANGE: Gila River basin in New Mexico and Arizona, southwards through the coastal rivers of Sonora
OVERALL SIZE: Males up to 5 cm (2 in); females up to 6 cm (2.4 in)
WATER REQUIREMENTS, DIET AND BREEDING: No detailed information regarding aquarium upkeep and breeding is available, but water conditions similar to those outlined for *Poeciliopsis gracilis*
NOTES: This very attractive species is officially regarded as threatened by the Endangered Species Committee of the American Fisheries Society. Captive-breeding is, obviously, a priority but care must be exercised to keep populations from different localities separate. These populations, which include some all-female assemblages, are regarded by some workers as subspecies – *P. o. occidentalis* and *P. o. sonoriensis*

4. *Poeciliopsis scarlli*
COMMON NAME: The Flier
RANGE: Freshwater canals near the coast from the Guerrero/Michoacán state boundary northwards to near Playa Azul, Mexico
OVERALL SIZE: Males around 3.5 cm (1.4 in); females around 4.5 cm (1.8 in)
WATER REQUIREMENTS: Neutral to slightly alkaline, soft to slightly hard water. Temperature: 22–24 °C (72–75 °F)
PREFERRED DIET: Live and dry foods
BREEDING: Small batches of fry (4 or 5) produced every few days over a period of about two weeks

Plate 121   *Poeciliopsis gracilis* pair (male above), from Veracruz in Mexico.
(*Manfred Meyer*)

Plate 122   *Poeciliopsis occidentalis* pair (male below), collected in Mexico.
(*Manfred Meyer*)

NOTES: The common name of this species relates to the angle at which the pectoral fins are often held by males, i.e. as if they are about to take off. The scientific name was erected by Meyer, Riehl, Dibble and me in 1985 (*Revue Française Aquariologie*, 12, 1, September 1985) in honour of John Scarll from Doncaster, Yorkshire, England, who, with UK aquarists Dennis Barrett and Dave Thompson, first collected the fish in 1984

5. OTHER *Poeciliopsis* SPECIES

A. Subgenus: *Poeciliopsis*

Poeciliopsis balsas Pacific slope of Mexico, including Río Balsas basin, Ríos Arteaga and Anguililla in Michoacán, Mexico

Poeciliopsis catemaco Laguna Catemaco, Veracruz, Mexico

Poeciliopsis hnilickai Ixtapa in Chiapas, Mexico

Poeciliopsis infans Highlands of the Río Grande de Santiago and Río Ameca basins in Jalisco and Michoacán, Mexico

Poeciliopsis latidens Pacific coast of Mexico from Río del Fuerte, southern Sonora, to near San Blas in Nayarit

Poeciliopsis lucida From northwestern Sinaloa to southeastern Sonora

Poeciliopsis monacha Restricted distribution in the vicinity of Rancho Guirocoba in southeastern Sonora, Mexico

Poeciliopsis presidionis Río Sinaloa to near San Blas in Nayarit – also found along the Pacific drainage of Mexico

Poeciliopsis prolifica Lower reaches of Río Yaquí in Sonora, southwards along the coast to San Blas in Nayarit, Mexico

Poeciliopsis turneri Río Purificación basin, southwards to Jalisco, Río Resolana to Cihuatlán

*Poeciliopsis turrubarensis** Pacific coast from Jalisco in Mexico to Río Dagua in Colombia

Poeciliopsis viriosa Río Ameca basin in Jalisco and southern Nayarit to Río Mocorito in Sinaloa, Mexico

Plate 123 *Poeciliopsis scarlli* – the Flier. This is a male first-generation aquarium specimen. (*Dennis Barrett*)

B. Subgenus: *Aulophallus*
Poeciliopsis elongata Pacific coast
of Costa Rica and western and
central Panama
Poeciliopsis   paucimaculata Río
General in Costa Rica
Poeciliopsis retropinna Costa Rica
and Río Chiriquí del Tire in
Panama

Of the above, only *P. infans*, *P. latidens*
and *P. viriosa* are encountered with
any regularity in the aquarium
hobby. Attempts are also currently
being made to establish captive-bred
populations of *P. elongata*

* A further species, *P. maldonadoi*, is
known from fossil remains only. It is very
similar in gonopodial characteristics to *P.
turrubarensis* and is probably conspecific
with it

GENUS: *Priapichthys*: Up to 1985 this
genus consisted of a number of species, including *Priapichthys annectens*, *P. festae*, *P. chocoensis*, and *P.
nigroventralis*. Following Radda's
revision (see *Pseudopoecilia* for full
reference), only *P. annectens* was
retained as a *Priapichthys*. The main
distinguishing characteristic of the
gonopodium is that ray 5a falls short
of the tip of the underlying ray 4p in
all species except *P. annectens*.
Differences in the gonopodial sus-
pensorium (the bones that support
the gonopodium and which are
located inside the body) also separate
P. annectens from its closest relatives. On a more easily detectable
level, *P. annectens* looks superficially
like a *Brachyraphis*, but with a longer
gonopodium

1.   *Priapichthys annectens*
SYNONYMS: *Priapichthys   annectens
annectens**, *P. a. hesperis**, *Gambu-
sia annectens*
RANGE: Atlantic and Pacific slopes of
Costa Rica
OVERALL SIZE: Males up to 5 cm (2 in);
females up to 8.5 cm (3.4 in)

WATER REQUIREMENTS: Neutral, softish
water with a little variation on either
side. Temperature range: 20–27 °C
(68–79 °F)
PREFERRED DIET: Live and dry foods
BREEDING: Around 30 fry produced
every 5–6 weeks
NOTES: * These two former subspecies
are now regarded as morphs of a
single variable species, *P. annectens*,
with the former *P. annectens annectens* being found on the Atlantic
slope of Costa Rica, and the former
P. a. hesperis on the Pacific side

GENUS: *Pseudopoecilia*: In 1985 Alfred
Radda resurrected this Regan 1913
generic name in his paper: 'Revision
der Gattung Priapichthys Regan
1913; sensu Rosen und Bailey
(1963)'; *Aquaria*, 32, pp 119–125. As
a result, *Priapichthys annectens*
remains within the same genus, while
all the others are transferred into
Pseudopoecilia which is characterised, among other parameters, by the
'tip of ray 5a falling short of tip of
underlying ray 4p which it touches'
(Rosen and Bailey, 1963)

1.   *Pseudopoecilia festae*
SYNONYMS: *Priapichthys festae*, *Poecilia
festae*, *Poecilia fria*, *Pseudopoecilia
fria*, *Priapichthys fria*
RANGE: Around Río Chico Chaune in
western Ecuador
OVERALL SIZE: Males around 3.5 cm
(1.4 in); females around 4.5 cm
(1.8 in)
WATER REQUIREMENTS: Alkaline, moderately hard to hard water. Temperature range 20–30 °C (68–86 °F), with
25 °C (77 °F) being a happy medium
PREFERRED DIET: Live and dry foods
BREEDING: About 20–25 fry produced
every 4–6 weeks
NOTES: This is a particularly attractive
fish when kept in appropriate water
conditions. Several good captive-
bred populations are available in
West Germany, the UK and the US

Plate 124 *Priapichthys annectens* (male). (*Dennis Barrett*)

Plate 125 *Pseudopoecilia festae* – a first-generation male. (*Dennis Barrett*)

Plate 126 *Pseudopoecilia nigroventralis* (male), from Anelagoya in Colombia. (*Manfred Meyer*)

2. *Pseudopoecilia nigroventralis*
SYNONYMS: *Priapichthys nigroventralis, Gambusia nigroventralis, G. caudovittata, Alloheterandria nigroventralis*
RANGE: Río Atrato and Río San Juán system, Colombia
OVERALL SIZE: Males about 2.5 cm (1 in); females about 3 cm (1.2 in)
WATER REQUIREMENTS AND DIET: Generally as for *P. festae*
BREEDING: Small broods – no accurate details yet available. Gestation period: around 6 weeks
NOTES: This striking species is a relative newcomer to the hobby. Exact details of its general upkeep are therefore not yet known. However, captive-bred populations are being currently established in a number of countries

3. OTHER *Pseudopoecilia* SPECIES
The following species are also known, the most recent one, *P. austrocolumbiana*, being described by Radda in 1988.
Pseudopoecilia austrocolumbiana Río Mira and Río Patia/Telembi systems in southern Colombia
Pseudopoecilia caliensis Cali in Colombia and Alto Cauca, Distrito de Bolo and La Quebrada de las Cruces
Pseudopoecilia chocoensis Río Calima, a tributary of the Río San Juán in the Pacific drainage of Chocó, Colombia
Pseudopoecilia dariensis Pacific slope of the eastern half of Panama
*Pseudopoecilia panamensis** Pacific drainage of Panama

* This is regarded as a synonym of *P. dariensis* by Radda (1985).

GENUS: *Xenophallus*: The single species in this genus, *X. umbratilis*, can be distinguished from *Neoheterandria* because it has an asymmetrical gono-

podium, some specimens exhibiting dextral (right-handed) asymmetry, others sinistral (left-handed) asymmetry.

1. *Xenophallus umbratilis*
SYNONYMS: *Neoheterandria umbratilis, Gambusia umbratilis, Brachyraphis umbratilis, Poeciliopsis maculifer*
RANGE: Atlantic drainage of Costa Rica
OVERALL SIZE: Males up to 5 cm (2 in); females up to 6.5 cm (2.6 in)
WATER REQUIREMENTS: Neutral to slightly alkaline, soft to slightly hard water. Temperature: 22–26 °C (72–79 °F)
PREFERRED DIET: Live and dry foods
BREEDING: Around 25 fry produced every 4–6 weeks
NOTES: This species is still quite often referred to as *Neoheterandria umbratilis*. However, the distinct asymmetry of the gonopodium would appear to support its separation

TRIBE: XENODEXIINI

GENUS: *Xenodexia*: The single species in this genus and tribe is *Xenodexia ctenolepis*, a unique Poeciliine in that it does not possess the usual gonopodial serrae, spines, claws, etc., associated with all the other members in the subfamily. The tube-like gonopodium exhibits dextral (right-handed) asymmetry. Further, the right pectoral fin in males has an accessory structure called a clasper. The exact use of this clasper has not yet been fully determined but it is assumed to be connected with the accurate direction, or stabilisation, of the gonopodium during copulation

1. *Xenodexia ctenolepis*
RANGE: Centred around the Río Ixcan in central Guatemala
OVERALL SIZE: About 4.5 cm (1.8 in) for both sexes
WATER REQUIREMENTS: Vigorously aerated water. Temperature range on

Plate 127 *Xenophallus umbratilis* pair (male above), from Costa Rica. (*Manfred Meyer*)

Plate 128 *Xenodexia ctenolepis* (male), collected in Guatemala (Río Ixcan). (*Ross Socolof*)

the cool side, from around 15.5°C
(60°F) to 25°C (75°F). Chemical
conditions are not critical but alkaline, hardish water is probably preferred (based on the Florida water in
which the 1989 collection was being
kept at the time of writing*)

PREFERRED DIET: Live and dry foods

BREEDING: So far, very small broods of a
maximum of 6 fry, produced over a
period of 4–5 days, followed by
a break, of about 3 months, have
been reported (Jaap-Jan de Greef,
personal communication) but these
figures may be found to be atypical of
the species once it becomes better
known within the hobby

NOTES: * The most recent collection of
this unique species was made by Ross
Socolof and Jaap-Jan de Greef in
early 1989. About 100 specimens
were collected, most of which survived the trip back to Florida. It is
hoped that several captive-bred
populations will become established
in the near future as a result of this
collection

Mature males develop a very small
sword on the lower rays of the caudal
fin, and some females also exhibit
this characteristic (Jaap-Jan de
Greef, 'Aquarium Care and Breeding of *Xenodexia ctenolepis*', *Aquarist & Pondkeeper*, April 1991)

FAMILY: **ANABLEPIDAE**
SUBFAMILY: ANABLEPINAE

GENUS: *Anableps*: This genus is com-
posed of the distinctive Four-eyed
Fishes in which each eye (there are
only two, of course!) is divided in
such a way as to allow visibility above
and below the water simultaneously.
Some of the rays of the tube-like
gonopodium (which is twisted either
to the left or the right) are themselves
twisted around each other

1.   *Anableps anableps*

SYNONYMS: *Anableps anonymus, A. gro-
novii, A. lineatus, A. surinamensis,
A. tetrophthalmus*

COMMON NAMES: Four-eyed   Fish,
Four-eyes

Plate 129   *Anableps anableps*, the Four-eyed Fish. (*Bill Tomey*)

Plate 130   The One-sided Livebearer – *Jenynsia lineata* (male). (*Wilf Blundell*)

RANGE: Brackish and fresh waters from southern Mexico southwards into northern South America

OVERALL SIZE: Males from around 15 cm (6 in); females up to 30 cm (12 in)

WATER REQUIREMENTS: Shallow water – 20–30 cm (8–12 in) – with ample clear swimming areas and plenty of space above it. Brackish conditions – approximately 5 ml (1 teaspoonful) of salt per 4.5 litres (1 Imperial gallon). Temperature: 22–30 °C (72–86 °F)

PREFERRED DIET: Floating livefoods (including any aerial insects that fall in), plus floating dry foods

BREEDING: Large – 3–4 cm (1.2–1.6 in) – fry produced. While 13 fry have been reported (Pat and Tom Bridges), more normal broods number less than six fry produced in two broods per year

NOTES: *Anableps* species are excellent jumpers so a good aquarium cover is a must. Specimens may come out of the water to feed if a suitable means of doing so is provided (see Aquarium Layout section in Part II)

2.   OTHER *Anableps* SPECIES

Two other species, *A. dowi* (from the Pacific coastal areas of central America) and *A. microlepis* (from the Atlantic side), both having similar characteristics and requirements to *A. anableps*, are occasionally available

GENUS: *Jenynsia*: This monotypic (single-species) genus is distinguished from the other Anablepinae in having an unscaled tubular gonopodium. As in *Anableps*, males can only deflect the gonopodium either to the right or the left. Females have similarly deflected genital apertures

1.   *Jenynsia lineata*

SYNONYMS: *Lebias lineata, L. multidentata, Poecilia punctata, Cyprinodon lineatus, C. multidentatus, Fitzroyia multidentatus, Xiphophorus heckeli, Jenynsia eigenmanni*, J. pygogramma*, J. maculata*, J. multidentata** and others

COMMON NAME: One-sided Livebearer

RANGE: South of the Amazon from southern Brazil southwards through Uruguay to the Río de la Plata in Argentina

OVERALL SIZE: Males around 4 cm (1.6 in); females around 12 cm (4.7 in)

WATER REQUIREMENTS: Well-aerated water with some salt – approximately 5 ml (1 teaspoonful) per 4.5 litres (1 gallon). Will also tolerate fresh water. Wide temperature range: from as low as 12 °C (54 °F) – for a short time – to 25 °C (77 °F). I have, however, collected *Jenynsia* in the polluted brackish Lago Rodrigo de Freitas in Rio de Janeiro in water whose temperature was 34.4°C (94°F)

PREFERRED DIET: Live and dry foods, including a vegetable component

BREEDING: Broods of around 20 very large fry (nourished by the mother, following ovulation) produced every 5–7 weeks

NOTES: Because of the deflected genital organs, right-handed males can mate only with left-handed females, and vice versa. A strain of *Jenynsia* that inhabits high-salinity lagoons has also been reported

* These fish are regarded as distinct species by Lynne Parenti in 'A Phylogenetic and Biogeographic Analysis of Cyprinodontiform Fishes (Teleostei, Atherinomorpha)' (*Bulletin of the American Museum of Natural History*, Vol. 168, Article 4, 1981)

FAMILY: **GOODEIDAE**
SUBFAMILY: GOODEINAE

GENUS: *Allodontichthys*: This is a genus of relatively small elongate fish with rounded heads and well developed pectoral fins. The teeth are different to those of other genera in being 'keeled' on either edge of their anterior face

1. *Allodontichthys tamazulae*

RANGE: Ríos Terrero and Tamazula in Jalisco, Mexico

OVERALL SIZE: Both sexes around 4.5 cm (1.8 in)

WATER REQUIREMENTS: Slightly to moderately alkaline water. Temperature: 18–24 °C (65–75 °F)

PREFERRED DIET: Live and dry foods, plus a vegetable component

BREEDING: Very small broods of 5 or fewer fry produced in aquaria so far. Long gestation period of 7–8 weeks or longer

NOTES: This species looks superficially like an *Ilyodon* but is less elongate and considerably smaller

OTHER *Allodontichthys* SPECIES

Three other species of *Allodontichthys*, all with basically similar requirements, are recognised: *A. hubbsi* (from the Río Tuxpan drainage, Río Terrero and Río Coahuayana in Jalisco, Mexico); *A. zonistius* (from Río Colima in Colima, Mexico); and, most recently, *A. polylepis*

GENUS: *Alloophorus*: This genus is distinguished, among other things, by possessing a large head and robust-looking body

1. *Alloophorus robustus*

SYNONYMS: *Fundulus robustus*, *F. parvipinnis*, *Zoogoneticus robustus*, *Z. maculatus*

RANGE: Río Lerma basin (Río Grande de Santiago) in Jalisco, Mexico

OVERALL SIZE: Males around 10 cm (4 in); females around 12 cm (4.7 in)

WATER REQUIREMENTS AND DIET: Generally as for *Allodontichthys tamazulae*

BREEDING: Large broods – up to 64 fry reported by Dr Joanne Norton. Gestation period: around 8 weeks

NOTES: This can be an aggressive, cannibalistic fish, so it is best kept in a species tank. Females should be separated from their fry. Another, smaller, species, *A. regalis*, from Los Reyes in Michoacán, Mexico, is also known but hardly ever kept

Plate 131 *Allodontichthys tamazulae* (male), from Río Terrero in Jalisco, Mexico. (*Manfred Meyer*)

Plate 132 *Alloophorus robustus* (female), from the Río Lerma basin in Mexico. (*Manfred Meyer*)

GENUS: *Allotoca:* Females of this genus have a tendency to become very thick-bodied when gravid and can become somewhat misshapen. The trophotaeniae (embryo-nourishing structures) have two anterior and two posterior arms. The dorsal fin is located very far back on the body.

1.   *Allotoca catarinae* *

SYNONYMS: *Neoophorus catarinae, Neoophorus diazi, N. d. catarinae*

RANGE: Laguna Santa Catarina, in Michoacán, Mexico

OVERALL SIZE: Males round 5 cm (2 in); females up to 8 cm (3.2 in)

WATER REQUIREMENTS AND DIET: Generally as for *Allodontichthys tamazulae*

BREEDING: About 10–15 fry reported. Long gestation period of around 7–8 weeks

NOTES: * *A. catarinae* was formerly regarded as a *Neoophorus* species but was transferred into *Allotoca* by Michael Leonard Smith and Robert Rush Miller in 1987 in their paper: '*Allotoca goslinae*, a New Species of Goodeid fish from Jalisco, Mexico'

(*Copeia*, No. 3). Accordingly, the genus *Neoophorus* should, if the above becomes universally accepted, cease to exist

2.   *Allotoca dugesi*

SYNONYMS: *Fundulus dugèsii, Adinia dugesii, Zoogoneticus dugèsii, Allotoca dugesii, A. vivipara*

COMMON NAME: Golden Bumblebee Goodeid

RANGE: Laguna de San Marcos (Río Verde) in the Río Lerma basin in Jalisco; Lagos de Cuitzeo, Pátzcuaro and Zirahuén in Michoacán, and Guanajuato, Mexico

OVERALL SIZE: Males around 6 cm (2.4 in); females slightly larger

WATER REQUIREMENTS AND DIET: Generally as for *Allodontichthys tamazulae*

BREEDING: Up to 76 fry reported by Derek Lambert

NOTES: Males and females are very differently pigmented in this species (both are very attractive). *A. dugesi* can be quite aggressive and is best kept in a species tank

Plate 133   *Allotoca catarinae* (male), collected in Laguna Santa Catarina in Mexico. (*Manfred Meyer*)

Plate 134 *Allotoca dugesi*, the Golden Bumblebee Goodeid (male.) (*Derek Lambert*)

Plate 135 *Allotoca dugesi* females do not possess the golden colours found in males. (*Derek Lambert*)

3. OTHER *Allotoca* SPECIES

Four other species, *A. diazi* (formerly *Neoophorus diazi*) (from Lago de Pátzcuaro in Michoacán, Mexico), *A. (Neoophorus) meeki* (from Lago Zirahuén in Michoacán), *A. maculata* (from Laguna de Santa Magdalena and Etzatlán in Jalisco, Mexico), and *A. goslinae*, also from Jalisco, are occasionally available as well

GENUS: *Ameca: Ameca splendens*, the single species in this genus, is a more deep-bodied fish than any of the previous ones. In addition, males, in particular, have a very full dorsal fin (hence the common name for the species) and an extremely elongated caudal peduncle

1. *Ameca splendens*

COMMON NAMES: Butterfly Goodeid; Ameca

RANGE: Ríos Ameca and Teuchitlán in the Río Ameca basin, Jalisco, Mexico

OVERALL SIZE: Males up to about 8 cm (3.2 in); females up to 12 cm (4.7 in) but usually smaller

WATER REQUIREMENTS: Not critical but alkaline, medium-hard conditions seem to be preferred. Wide temperature range: 20–29 °C (70–84 °F)

PREFERRED DIET: All types of food, but must include a vegetable component

BREEDING: Broods of over 40 fry are quite common in older females. Young females can produce as few as 6 fry, often of widely varying sizes and (usually in the first-ever brood) with slightly misshapen bodies. Gestation period up to 8 weeks or longer at cool temperatures

NOTES: This species, more than any other, was responsible for bringing Goodeids to the notice of European aquarists in the early 1970s

GENUS: *Ataeniobius*: The single species in this genus, *A. toweri*, is a slim, elongated fish which has traditionally been reported as being the only Goodeid that does not possess trophotaeniae. This statement should, however, be treated with some caution, since scanning electron micrographs of new-born fry which I took some time ago indicate that there appears to be some trophotaenial development. This finding is supported by observations of late embryos made by the British aquarist Derek Lambert

1. *Ataeniobius toweri*

SYNONYM: *Goodea toweri*

COMMON NAME: Blue-tailed Goodeid

RANGE: Río Verde, San Luis Potosí, Mexico

OVERALL SIZE: Males around 6 cm (2.4 in); females around 7 cm (2.8 in)

WATER REQUIREMENTS AND DIET: Generally as for *Ameca splendens*, avoiding lower end of the temperature scale

BREEDING: Up to 37 fry reported by UK aquarist Ivan Dibble. Gestation period about 7–8 weeks, with a seasonal rest during the cooler months

NOTES: Males of this attractively bluish-coloured species will swim above receptive females during courtship, touching them on the top of their head with their chin. Research being undertaken in the US at the time of writing may lead to the eventual transfer of this species to the genus *Goodea*

GENUS: *Chapalichthys*: Members of this genus have relatively deep bodies, smallish heads and (in males) large dorsal fins similar in shape and size to those of *Ameca splendens* and *Xenoophorus captivus*. Males also have a yellow vertical band in the caudal fin reminiscent of that in *A. splendens*

1. *Chapalichthys encaustus*

SYNONYM: *Characodon encaustus*

COMMON NAME: Spotted Goodeid

RANGE: Laguna de Chapala and Río Grande de Santiago in Jalisco; Río Tanhuato and Lago Cuitzeo in Michoacán, Mexico

Plate 136 *Ameca splendens* – the Butterfly Goodeid. This specimen is an adult male from the Río Ameca basin in Jalisco, Mexico. (*Manfred Meyer*)

Plate 137 *Ataeniobius toweri* pair (male below). (*'Aquarian' Fish Foods*)

OVERALL SIZE: Males around 6 cm (2.4 in); females about 8 cm (3.2 in)

WATER REQUIREMENTS: Slightly to moderately alkaline, medium-hard water. Temperature range: 20–28 °C (68–82 °F)

PREFERRED DIET: Wide range of foods, including a vegetable component

BREEDING: About 50 young produced by large females. Gestation period: about 6–8 weeks

NOTES: The yellow caudal fin band that males exhibit when in peak condition will tend to fade or disappear completely if the pH is allowed to drop below neutral

2. Chapalichthys pardalis
RANGE: Tocumbo, Michoacán, Mexico

OVERALL SIZE: Males about 6 cm (2.4 in); females about 7 cm (2.8 in) or slightly larger

WATER REQUIREMENTS, DIET AND BREEDING: As for Chapalichthys encaustus

NOTES: The body coloration of both sexes is very similar to that of Ameca splendens females. Males have the characteristically yellow caudal fin band found in C. encaustus and A. splendens. The dorsal fin in males is also similar to that found in Ameca and Xenoophorus. The similarities between the genera could conceivably turn out, if examined scientifically, to be more than merely apparent. If so, it would then not seem to be outside the realm of possibility for all three genera to be synonymised at some stage in the future, with Chapalichthys, being the oldest generic name, taking priority. A third Chapalichthys species, C. peraticus, was imported into the UK by aquarist Ivan Dibble in 1982, but it has not, as yet, become established in captivity

GENUS: Characodon: Members of this genus are relatively small, with a robust-looking body, smallish rounded head and large eyes (particularly noticeable in small adults). The dorsal and anal fins are set well back

on the body. The embryos' trophotaeniae have two long posterior extensions but no anterior extensions of any consequence

1. Characodon audax
COMMON NAMES: Black Prince*, Bold Characodon

RANGE: Outflow of El Ojo de Agua de las Mujeres at El Toboso in Durango, Mexico

OVERALL SIZE: Males around 4.5 cm (1.8 in); females around 5 cm (2 in)

WATER REQUIREMENTS: Well-planted, quiet tank, with slightly to moderately alkaline, medium-hard water. Temperature range: 20–24 °C (68–75 °F). Regular partial water changes recommended

PREFERRED DIET: Predominantly livefood based, but will accept other foods as well, including a vegetable-based component

BREEDING: Small broods of around 10–15 fry produced every 8 weeks or so, but as many as 59 have been reported (Derek Lambert)

NOTES: This is a shy, retiring species which dislikes turbulent water conditions and energetic tankmates. Owing to its very restricted range, C. audax must be considered as potentially under threat in the wild. There are, however, aquarium stocks of this species, so its future may be somewhat more secure than its natural distribution might indicate

* This most appropriate common name was coined by UK aquarist Ivan Dibble

2. Characodon lateralis
SYNONYM: Characodon garmani*

RANGE: Widespread (but not abundant) in springs, creeks and ponds in the upper Río Mezquital in Durango, Mexico

OVERALL SIZE: Males around 4 cm (1.6 in); females 5.5 cm (2.2 in)

WATER REQUIREMENTS, DIET AND BREEDING: Generally as for C. audax, but largest brood reported is only 27 (Ginny Eckstein)

Plate 138   *Chapalichthys encaustus* (male), from Lake Chapala, Mexico. (*Manfred Meyer*)

Plate 139   *Chapalichthys pardalis* (male) – aquarium specimen. (*Dennis Barrett*)

NOTES: This species will hybridise with *C. audax*. Both species should therefore be kept well away from each other

* *C. garmani* is afforded full specific status by Smith and Miller 'Mexican Goodeid Fishes of the Genus *Characodon*, with Description of a New Species' (*American Museum Novitates*, No. 2851, pp 1–14, June 1986). They base their conclusion on the isolated nature of the *C. garmani* population from Parras in Coahuila, Mexico and the distinct morphometric features of the holotype (a female). No other specimens are known and *C. garmani* is therefore presumed to be extinct

GENUS: *Girardinichthys*: This genus is represented by two smallish species, both of which possess a compressed body form (more noticeable in males). The dorsal and anal fins are broad-based and large in comparison to many other Goodeids. The trophotaeniae (embryo-nourishing structures) have two long posterior extensions and two short anterior extensions

1. *Girardinichthys multiradiatus*

SYNONYMS: *Girardinichthys innominatus*, *G. limnurgus*, *Characodon multiradiatus*, *Lermichthys multiradiatus*

COMMON NAME: Striped Goodeid

RANGE: Río and Lago de Lerma in the State of Mexico (i.e. around Mexico City) and Laguna Zempoala in Morelos, Mexico

OVERALL SIZE: Males around 3.5 cm (1.4 in); females around 5 cm (2 in)

WATER REQUIREMENTS: Well-planted tank containing alkaline, moderately hard water and little turbulence. Temperature range: 20–27 °C (68–80 °F) – though slightly higher temperatures will be tolerated

PREFERRED DIET: All foods, including a vegetable component

BREEDING: Up to 50 fry have been reported (Dr Joanne Norton) but broods are usually considerably smaller. Gestation period, dependent on temperature and health factors: around 7–8 weeks

NOTES: In the UK, this species has proved quite difficult to establish; US hobbyists have generally experienced a higher success rate

Plate 140   *Characodon audax* – the Black Prince. (*Dennis Barrett*)

Plate 141 *Characodon lateralis* pair (male below). (*'Aquarian' Fish Foods*)

Plate 142 *Girardinichthys multiradiatus* (male). (*Dennis Barrett*)

2. *Girardinichthys viviparus*

SYNONYMS: *Girardinichthys innominatus, Characodon geddesi, Lucania richi, Limnurgus innominatus, L. variegatus*

COMMON NAME: The Amarillo

RANGE: Xochimilco, Lago de Texcoco, Laguna de Tumpango, Chapultepec Park* in Mexico City

OVERALL SIZE: Males up to 5 cm (2 in), but usually smaller; females up to 7 cm (2.8 in)

WATER REQUIREMENTS: Generally as for *G. multiradiatus* but much cooler temperatures are tolerated. In fact, frost has been reported in the vicinity of the lake in Chapultepec Park where this species was formerly found

PREFERRED DIET: All foods, but must include a vegetable component ('green water' seems to be appropriate since the Chapultepec Park locality was heavily infested with free-floating green algae)

BREEDING: Broods generally around the 20 mark. Gestation period: 7–8 weeks

NOTES: * Recent attempts to collect this species from the lake in Chapultepec Park have proved fruitless. In the past, the Amarillo has been considered a difficult species in the UK. However, the latest batch of imports seems to have eased matters somewhat

GENUS: *Goodea*: There is some uncertainty regarding the correct classification of the various fish collectively referred to this genus. They are either recognised as full species in their own right or as in Meyer, Wischnath and Foerster, in *Lebendgebärende Zierfische* (Mergus, 1985) as subspecies of a single species, *Goodea atripinnis*. Research currently being carried out in the US should lead to clarification in the near future.

Goodea are large, elongated fish with large heads and with dorsal and anal fins set well to the rear of the body. The trophotaeniae (embryo-nourishing structures) are very small and rosette-like

1. *Goodea atripinnis*

SYNONYMS: *Goodea atripinnis atripinnis, G. caliente, G. calientis, Characodon atripinnis, C. variatus, Xenendum caliente*

COMMON NAME: Black-fin Goodeid

RANGE: León, Valle de Santiago, Alberca; Guanajuato; Río Ameca; Jalisco, Mexico

OVERALL SIZE: Males around 12 cm (4.7 in); females around 20 cm (7.9 in), usually smaller

WATER REQUIREMENTS: Alkaline, moderately hard water. Temperature range: 18–24 °C (64–75 °F)

PREFERRED DIET: Live and dry foods, with a vegetable component

BREEDING: About 50 fry per brood. Gestation period: 6–8 weeks

NOTES: Owing to its large size (and occasional fin-nipping tendencies), *G. atripinnis* should be kept in a species tank

2. OTHER Goodeas

Pending a review of the genus, the following three subspecies are recognised by various researchers in the field, including Meyer, Wischnath and Foerster (whose distribution data are used here):

Goodea atripinnis gracilis Río Santa Marina; San Luis Potosí; Río San Juán del Río; Querétaro, Mexico

Goodea atripinnis luitpoldi (Based on Hubbs and Turner, 1939) – Río Grande de Santiago, between Ocotlán and Laguna de Chapala in Jalisco, Mexico

Goodea atripinnis martini Río de Morelia, Lago de Quitzeo; Michoacán, Mexico

Goodea atripinnis xaliscone Laguna Chapala; Jalisco, Mexico

Plate 143 *Girardinichthys viviparus* pair (male above). (*Dennis Barrett*)

Plate 144 *Goodea atripinnis* (female), from Jalisco, Mexico. (*Manfred Meyer*)

GENUS: *Hubbsina*: The sole species in this genus, *Hubbsina turneri*, is distinguished from all other Goodeids in having a large, spectacular, almost rectangular, dorsal fin. In addition, the body is compressed and the head is large

1. *Hubbsina turneri*
RANGE: Michoacán (Cointzio, near Morelia) and Guanajuato, Mexico
OVERALL SIZE: Males about 5 cm (2 in); females 6.5 cm (2.6 in)
WATER REQUIREMENTS AND BREEDING: Little detailed information is available about the culture of this species. The few stocks that were kept for a while by Pat and Derek Lambert were maintained in hard and alkaline water at around 25 °C (77 °F). The fish were fed on live foods. Dry foods were refused
BREEDING: No details are currently available
NOTES: New attempts at collecting new stocks from the wild may result in aquarium populations of this spectacular fish being established in the near future

GENUS: *Ilyodon*: Members of this genus are very elongate in body form. They are also extremely fast swimmers. Considerable controversy exists regarding the classification of the various species with, in my opinion, justifiable doubt being cast on their validity. For instance, *Ilyodon xantusi* was demoted in status to that of a broad-mouthed morph of *I. furcidens* by Turner, Grudzien, Adkissa and White in 'Evolutionary Genetics of Trophic Differentiation in Goodeid Fishes of the Genus *Ilyodon*' (*Environmental Biology of Fishes*, Vol. 9, No. 2, pp 159–177, 1983). Earlier, Alvarez, in 1970, had synonymised *I. xantusi* with *I. whitei* (*Mexican Fishes* – Secretaria de Industria y Comercio, Comisión Consultativa de Pescas, Mexico City, pp 166). Then, Bruce Turner (*Tropical Fish Hobbyist*, July

1989) refers to *I. lennoni* as appearing 'to be doubtfully distinct from *I. whitei*'. Therefore, should the above be universally accepted in due course – and there seem to be very strong arguments for doing so – all four nominal species would end up being no more than different forms of a single species which (because it was described before all the others) would be *I. furcidens*.
Swayed by Turner's strong arguments, I am therefore tentatively subsuming all four *Ilyodon* 'species' under the *I. furcidens* banner, and would urge all interested readers to consult Bruce Turner's three comprehensive features for fuller details: 'The *Ilyodon* Story', *Tropical Fish Hobbyist*, July, August and September 1989

1. *Ilyodon furcidens**
SYNONYMS: *Characodon furcidens*, *Ilyodon paraguayense*, *I. lennoni**, *I. whitei**, *I. xantusi** and others
RANGE: Mainly in rivers draining the southern slope of the Mesa Central (Mexico) towards the Pacific Ocean
OVERALL SIZE: Males and females around 10 cm (4 in)
WATER REQUIREMENTS: Neutral to alkaline, slightly to moderately hard, well-aerated water. Temperature range: from around 22 °C (72 °F) to 27 °C (80 °F)
PREFERRED DIET: Live and dry foods, including a regular vegetable component
BREEDING: Generally between 20 and 35 fry (but over 50 have been reported). Gestation period: 7–8 weeks
NOTES: These highly mobile colourful fish develop a bit of a neck hump as they grow older

* *I. furcidens* and *I. 'xantusi'* are found together in Río Armeria and Coahuayana (and minor coastal rivers), while *I. 'lennoni'* and *I. 'whitei'* are from the Río Balsas system. A further 'new' *Ilyodon*, sometimes referred to as *I. 'amecae'*, from the Río Pola may also turn out to be another morph of this highly variable species

Plate 145   *Hubbsina turneri* (female), collected in Laguna de Zacapu, Michoacán, Mexico. (*Derek Lambert*)

Plate 146   *Ilyodon furcidens* (male), from Colima, Mexico. (*Manfred Meyer*)

Plate 147 *Ilyodon 'lennoni'* (male), from Río Chacamero in Guerrero, Mexico. (*Manfred Meyer*)

Plate 148 *Ilyodon 'whitei'* (male), from Cuautla, Mexico. (*Manfred Meyer*)

Plate 149 *Ilyodon 'xantusi'* (male) – aquarium specimen (*Wilf Blundell*)

GENUS: *Skiffia:* Members of this genus are small with compressed bodies and small heads. They can be easily distinguished from all other Goodeids in that males possess irregularly shaped dorsal fins that give rise to the common name of Sawfin Goodeids

1. *Skiffia bilineata*

SYNONYMS: *Characodon bilineata, Goodea bilineata, Neotoca bilineata*

COMMON NAMES: Two-lined Skiffia, Black-finned Goodeid, The Elfin

RANGE: Río Grande de Santiago in Jalisco, Michoacán and Guanajuato, and in the Río Lerma in Guanajuato, Mexico

OVERALL SIZE: Males up to 4 cm (1.6 in); females up to 6 cm (2.4 in)

WATER REQUIREMENTS: Well-planted aquaria. Slightly to moderately alkaline, hard water. Temperature range 22–25 °C (72–77 °F)

PREFERRED DIET: Live and dry foods

BREEDING: Up to 40 fry. Gestation period: around 8 weeks

NOTES: The black fins indicated by one of the common names for this species are well developed only in fully mature males

2. *Skiffia francesae*

COMMON NAME: Gold Sailfin Goodeid

RANGE: Río Teuchitlán in Jalisco, Mexico*

OVERALL SIZE: Males around 4.5 cm (1.8 in); females around 5 cm (2 in)

WATER REQUIREMENTS: Slightly to moderately alkaline, hard water. Temperature range: 22–25 °C (72–77 °F)

PREFERRED DIET: Live and dry foods

BREEDING: Broods of up to 36 fry have been reported (Ed Marcisz), with an interval of about 8 weeks between broods

NOTES: * *Skiffia francesae* appears to have been driven into extinction in the wild within a few years of its official scientific description (Dolores Irene Kingston in *Copeia*, No. 3, 1978) through the introduction of Red Platies (*Xiphophorus maculatus*) into the Río Teuchitlán. Several recent attempts to re-discover the species have resulted in failure. Fortunately, aquarium populations were established following a 1976 collection from the wild. These are currently being maintained in the UK, West Germany and the US.

S. francesae seems to be at its best –

Plate 150   *Skiffia bilineata* (male) – aquarium specimen. (*Dennis Barrett*)

Plate 151   *Skiffia francesae* (male below). (*'Aquarian' Fish Foods*)

and reproduces with greatest success – when kept in shoals. A beautiful hybrid between *S. francesae* and the closely related *S. multipunctatus* has been developed by James Langhammer. The Black Beauty, as this hybrid is known, is fertile and must therefore be kept well away from both of the parental species

3. OTHER *Skiffia* SPECIES
Two other species of *Skiffia* are encountered within the hobby. Neither can be regarded as easy

Skiffia lermae Lago de Pátzcuaro, Celaya, Michoacán in Mexico
Skiffia multipunctatus Lago Camécuaro and Río Lerma basin in Michoacán and Jalisco, Mexico

GENUS: *Xenoophorus*: The single species in this genus, *Xenoophorus captivus*, is a relatively deep-bodied fish with a large dorsal fin, not dissimilar to that of *Ameca* and *Chapalichthys* males. The trophotaeniae (embryonourishing structures) tend to have one long, forked process, plus several shorter, thinner ones

1. *Xenoophorus captivus*
SYNONYMS: *Goodea atripinnis*, *G. captiva*, *Xenoophorus erro**, *X. exsul**
RANGE: 'Captivus' type – near Jesús María in the Río Panuco basin, San Luis Potosí, Mexico; 'Erro' type – Río Santa Maria (a tributary of the Río Panuco, San Luis Potosí; 'Exsul' type – Agua del Medio, isolated streams north of Río Santa María, San Luis Potosí
OVERALL SIZE: Males around 5 cm (2 in); females around 6 cm (2.4 in)
WATER REQUIREMENTS: Not critical, but slightly to moderately alkaline, hard water seems to be preferred. Temperature range 20–25 °C (68–77 °F)
PREFERRED DIET: Live and dry foods, with a vegetable component
BREEDING: Around 35 fry produced every 8 weeks or so

NOTES: * The three types of *X. captivus* were formerly accorded full specific status by Hubbs and Turner in 1937. Although some morphological differences do exist between them, these are not generally regarded as being sufficiently significant to warrant their separation

GENUS: *Xenotaenia*: This genus is represented by a single species, *Xenotaenia resolanae*. It is a robust-looking fish with a large rounded head and relatively small dorsal fin, compared, e.g., with *Ameca*, *Chapalichthys* and *Xenoophorus*

1. *Xenotaenia resolanae*
RANGE: Río Resolana near Autlan in Jalisco, Mexico
OVERALL SIZE: Males around 4 cm (1.6 in); females around 5 cm (2 in), sometimes a little larger
WATER REQUIREMENTS AND DIET: Generally as for *Xenoophorus captivus*
BREEDING: Broods numbering between 10 and 30 fry. Gestation period: around 7–8 weeks
NOTES: This fish sometimes develops finnipping tendencies and should therefore be kept in a species tank. Individual specimens suspected of being hybrids have occasionally been reported, but not conclusively confirmed

GENUS: *Xenotoca*: Members of this genus have a relatively compressed body form. Males sport a well-developed dorsal fin. The characteristic Goodeid anal fin notch is particularly well developed in this genus

1. *Xenotoca eiseni*
SYNONYMS: *Characodon eiseni*, *Xenotoca variata*
COMMON NAMES: Orange-, or Red-tailed Goodeid
RANGE: El Sacristán, Río San Leonel

Plate 152   *Xenoophorus captivus* (male above). (*'Aquarian' Fish Foods*)

Plate 153   Fully mature *Xenotaenia resolanae* male. (*Wilf Blundell*)

and Río Grande de Santiago in Nayarit, and Río Tamazula in Jalisco, Mexico

OVERALL SIZE: Males up to 6 cm (2.4 in); females around 7 cm (2.8 in)

WATER REQUIREMENTS: Tolerant of wide range of conditions but moderately hard, alkaline water is preferred. Temperature range: wide tolerance reported from 15 °C (59 °F) to 30 °C (86 °F), but something in the region of 25 °C (77 °F) is recommended

PREFERRED DIET: Live and dry foods, plus a vegetable component

BREEDING: Broods of around 30 are quite common, but as many as 89 have been reported (C. L. Michaels). Gestation period: 7–8 weeks

NOTES: As males get older, they develop a very thick-set body with a pronounced neck hump. This species also tends to develop fin-nipping tendencies with age

2. *Xenotoca variata*

SYNONYMS: *Characodon variatus*, *C. ferrugineus*

RANGE: Río Lerma basin in Guanajuato, Río Santa María in San Luis Potosí, Río Grande de Santiago and near Lago de Chapala in Jalisco and Michoacán, and Río de la Laja drainage in Querétaro – all in Mexico

OVERALL SIZE: Males around 6 cm (2.4 in); females around 7.5 cm (3 in)

WATER REQUIREMENTS AND DIET: Generally as for *X. eiseni*

BREEDING: Smaller broods than *X. eiseni* (averaging around 25), but as many as 39 fry have been reported (Bob Clarke). Gestation period: 7–8 weeks

NOTES: This species is not as deep-bodied as *X. eiseni* and bears some resemblance to *Ameca splendens* (even down to the yellow vertical band on the caudal fin in males)

Plate 154 *Xenotoca eiseni* (male), from Nayarit in Mexico. (*Manfred Meyer*)

Plate 155   *Xenotoca variata* – first-generation captive-bred male. (*John Dawes*)

Plate 156   *Zoogoneticus quitzeoensis* – yellow-finned aquarium-bred male. (*Wilf Blundell*)

3. OTHER *Xenotoca* SPECIES

A third species, *X. melanosoma*, with similar general requirements to *X. eiseni* and *X. variata*, is also in existence within the aquarium hobby. It is about the same size as *X. variata* but is considerably less colourful. *X. melanosoma* is found in Río Tamazula, Río Ameca and Río Grande de Santiago, all in Jalisco, Mexico

GENUS: *Zoogoneticus*: This is another monotypic genus of Goodeid (but see Notes). Its sole representative is *Zoogoneticus quitzeoensis*. It is a more delicate-looking species than most other Goodeids and is smaller than most. Males have a compressed body reminiscent of *Skiffia* males but lack the 'sawfin' dorsal

1. *Zoogoneticus quitzeoensis*

SYNONYM: *Platypoecilus quitzeoensis*

RANGE: Río Grande de Santiago in Jalisco, Lago de Zacapu and Lago Cuitzeo in Michoacán, Ojo de Agua de Santiaguillo in Guanajuato, Mexico

OVERALL SIZE: Males from (Yellow morph) 3.5 cm (1.4 in) to (Orange morph) around 5 cm (2 in); females from (Yellow morph) 4.5 cm (1.8 in) to (Orange morph) 7.6 cm (3 in)

WATER REQUIREMENTS: Slightly to moderately alkaline, hard water. Temperature range: 22–27 °C (72–79 °F)

PREFERRED DIET: Live and dry foods, including a vegetable component

BREEDING: This is not an easy species. Broods average 15 fry. Gestation period: around 8 weeks, but can vary considerably

NOTES: This is a very attractive, rather shy and slow-growing species. Three colour morphs are known: Yellow, the smallest, known largely from aquarium specimens; Red, intermediate in size, found in the area surrounding Lake Chapala, and Orange, the largest form, found in Lake Zacapu by UK aquarists Derek and Pat Lambert, along with several other Goodeid species, e.g. *Hubbsina turneri*, *Xenotoca variata*, *Goodea atripinnis* and *Allotoca dugesi*. Derek Lambert, Dr Michael, L. Smith, C. Rodriguez and L. Butler have recently collected a different *Zoogoneticus* which has been given the common name of Crescent Zoe, pending scientific investigation and description. See 'The Crescent Zoe' by Derek Lambert in *Aquarist & Pondkeeper* Vol. 55, No. 9. pp. 26–27 (December 1990)

FAMILY: **RIVULIDAE**

GENUS: *Cynolebias*: This genus of Killifish is distinguished from other Aplocheiloids in having a completely closed preopercular canal.

1. *Cynolebias brucei*

SYNONYM: *Campellolebias brucei*

COMMON NAME: Turner's Gaucho

RANGE: Mainly around Santa Catalina, Brazil

OVERALL SIZE: Males around 3 cm (1.2 in); females somewhat smaller

WATER REQUIREMENTS: Slightly acid, soft water. Temperature around 25 °C (77 °F)

PREFERRED DIET: Livefoods

BREEDING: This is an annual fish capable of laying eggs that have previously been fertilised internally. Incubation time, in moist peat, from several weeks to several months depending, largely, on temperature. Eggs require soaking in soft acid water when the embryos are seen to be fully eyed-up (developed)

NOTES: Along with *C. melanotaenia*, *C. brucei* is included in this book because it meets John Wourm's criteria as a facultative livebearer (see Part I for fuller details)

Plate 157   Lake Zacapu in Michoacán, Mexico. This locality is particularly rich in Goodeid species, containing – in addition to *Zoogoneticus quitzeoensis* – at least *Hubbsina turneri*, *Xenotoca variata*, *Goodea atripinnis* and *Allotoca dugesi*. (*Derek Lambert*)

Plate 158   *Dermogenys pusillus* – the Wrestling Halfbeak. (*Harry Grier/Florida Tropical Fish Farms Association*)

2. *Cynolebias melanotaenia*
SYNONYM: *Cynopoecilus melanotaenia*
COMMON NAME: Fighting Gaucho
RANGE: Around Porto Alegre, Brazil
OVERALL SIZE: Males around 5 cm (2 in); females smaller
WATER REQUIREMENTS, DIET AND BREEDING: Generally as for *C. brucei*
NOTES: This is a very hardy species. Two types of eggs can be laid – some smooth and some with short spines. Whether this is linked in any way with internal or external fertilisation does not appear to be known

FAMILY: **ORYZIIDAE (ORYZIATIDAE)** sensu Nelson and others ≡ Adrianichthyidae (sensu Rosen and Parenti)

GENUS: *Oryzias*: This genus is distinguished from other Adrianichthyoids (as defined by Nelson and others) in that the jaws are 'not tremendously enlarged'. While *Oryzias* are generally regarded as egglayers it has been noted by some workers that, in *O. latipes* at least, it is possible for eggs to be fertilised internally

1. *Oryzias latipes*
COMMON NAMES: Geisha Girl, Japanese Medaka, Rice Fish
RANGE: Japan
OVERALL SIZE: Both sexes around 4 cm (1.5 in)
WATER REQUIREMENTS: Neutral conditions with tolerance on either side. Temperature: extremely wide range, from 15 °C (59 °F) to 28 °C (82 °F)
PREFERRED DIET: Live and dry foods
BREEDING: Following either internal or external fertilisation, females carry the eggs attached to their genital apertures for a time after mating, later depositing them among vegetation. Hatching time: 10–12 days, depending on temperature

NOTES: Golden and red (unreliable) varieties of this species are also available. *Oryzias* a faculative livebearer as defined by John Wourms (1981) – see Part I for further discussion of this point

FAMILY: **HEMIRHAMPHIDAE**
SUBFAMILY: HEMIRHAMPHINAE

GENUS: *Dermogenys*: Members of this genus are all elongate fish with fins set well back on the body. The lower jaw is considerably longer than the upper one and bears no teeth

1. *Dermogenys pusillus* *
SYNONYM: *Hemirhamphus fluviatilis*
COMMON NAMES: Wrestling Halfbeak, Malayan Halfbeak
RANGE: Thailand, Malaysia, Java, Sumatra, Singapore, Kalimantan, Indonesia
OVERALL SIZE: Males up to 6 cm (2.4 in); females around 8 cm (3.2 in)
WATER REQUIREMENTS: Fresh or brackish water – 5 ml (1 teaspoonful) of salt per 4.5 litres (1 Imperial gallon). Temperature range: 20–30 °C (68–86 °F) or even higher
PREFERRED DIET: Floating foods consisting largely of livefoods
BREEDING: Up to 40 large – 1 cm (0.4 in) – fry produced by large females every 4–8 weeks
NOTES: As the first common name suggests, males can be quite aggressive towards each other, sometimes engaging in long, drawn out contests lasting 20 minutes or more

* Various subspecies are recognised by some authors, indicating the wide distribution of the species: *D. pusillus pusillus*, *D. p. burmanicus*, *D. p. orientalis*, *D. p. siamensis*, and *D. p. sumatranus*

2. OTHER *Dermogenys* SPECIES

Several other species are recognised, of which few are ever seen within the aquarium hobby.

Dermogenys megarrhamphus Towuti Danau (Lake Towoeti), central Sulawesi

Dermogenys montanus Waterfalls at Bantimurung, Sulawesi

Dermogenys philippinus Kulaman Plateau in Cebu, Philippines

Dermogenys viviparus Islands of Samar and Luzón in the Philippines

Dermogenys vogti Topobulu, Sulawesi

Dermogenys weberi Matana, Sulawesi

GENUS: *Hemirhamphodon:* Hemirhamphodons are elongate fish in which, in the male, the leading edge of the anal fin is extremely long, while the posterior edge is strongly truncated. A further distinguishing factor between this genus and *Dermogenys* is that *Hemirhamphodon* species possess teeth in their lower jaw

1. *Hemirhamphodon pogonognathus*

SYNONYMS: *Hemiramphus pogonognathus*, *Hemiramphodon kükenthali*

RANGE: Malaysia, Singapore, Sumatra, Kalimantan, Bangka, Belitung (Billiton), Halmahera

OVERALL SIZE: Both males and females around 8 cm (3.2 in)

WATER REQUIREMENTS, DIET AND BREEDING: Generally, as for *Dermogenys pusillus*

NOTES: This is a slender species with delicate but very attractive colours that can be appreciated fully only in well-maintained aquaria with good lighting

2. OTHER *Hemirhamphodon* SPECIES

Two other species of this genus are occasionally available:

Hemirhamphodon chrysopunctatus Mainly found around Bandjamarsin in southern Kalimantan

Hemirhamphodon phaiosoma Singapore, Belitung (Billiton), Bangka and Kalimantan

GENUS: *Nomorhamphus*: The species in this genus are generally thicker-bodied than those of the two preceeding genera and the lower jaw is not as extended. In addition, the tip of the lower jaw has a fleshy tip, particularly conspicuous in *N. liemi* males

1. *Nomorhamphus liemi**

COMMON NAME: Celebes Halfbeak

RANGE: Areas around Maros in southern Sulawesi

OVERALL SIZE: Males around 6 cm (2.4 in); females around 9 cm (3.5 in)

WATER REQUIREMENTS: Slightly acid to slightly alkaline, soft to medium-hard water. Temperature range: 24–27 °C (75–80 °F)

PREFERRED DIET: As for *Dermogenys pusillus*

BREEDING: Small broods of around 10 large – 1.8 cm (0.7 in) – fry. Gestation period: 6–8 weeks

NOTES: Older males exhibit a black recurved fleshy tip on their lower jaw

* Two different types of *N. liemi* are recognised, usually afforded subspecific status by most writers: *N. liemi liemi* is the type normally referred to as the Celebes Halfbeak and is recognised by its reddish fins edged in black; *N. liemi snijdersi*, found east of Maros, has much more extensive black pigmentation on its fins.

2. OTHER *Nomorhamphus* SPECIES

Several other species of *Nomorhamphus* are occasionally available, one of which, *N. celebensis*, rather than *N. liemi*, should perhaps be *the* Celebes Halfbeak.

Nomorhamphus brembachi southern Sulawesi

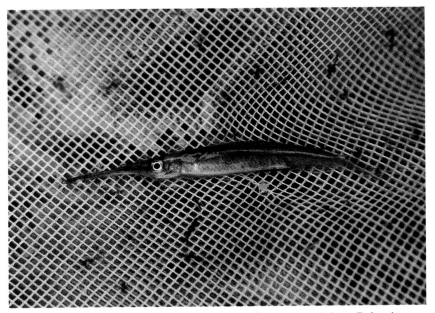

Plate 159 *Hemirhamphodon pogonognathus*, collected near Johore Bahru in southern Malaysia. (*John Dawes*)

Plate 160 *Nomorhamphus liemi* – Celebes Halfbeak male of the type usually assigned the subspecific name *N. liemi liemi. (Dennis Barrett)*

Nomorhamphus celebensis Poso Danau (Lake Posso), central Sulawesi

Nomorhamphus ebrardti Kabajana (Kabaena), Penango and Rumbia-Ebene in southeastern Sulawesi

Nomorhamphus hageni Penango and Rumbia-Ebene in southeastern Sulawesi

Nomorhamphus towoetii Towuti Danau (Lake Towoeti), central Sulawesi

FAMILY: **CHARACIDAE** (sensu Géry) ≡ Subfamily Characinae (sensu Nelson and others)
SUBFAMILY: GLANDULOCAUDI-NAE (Géry)
TRIBE: GLANDULOCAUDINI

GENUS: *Corynopoma*: This genus is distinguished from other members of the tribe (commonly referred to as the Croaking Tetras) by the possession of a pen-like appendage with a paddle-like tip extending from the posterior edge of the gill covers of mature males, the arrangement of the lower rays of the caudal fin into a spur or 'tail' and the absence of an adipose fin

1.   *Corynopoma riisei*
SYNONYMS: *Corynopoma albipinne, C. aliata, C. searlesi, C. veedoni, Nematopoma searlesi, Stevardia albipinnis, S. aliata, S. riisei*
COMMON NAME: Swordtail Characin
RANGE: Río Meta in Colombia
OVERALL SIZE: Both sexes around 7 cm (2.8 in)
WATER REQUIREMENTS: Slightly acid to slightly alkaline water of low to medium hardness. Temperature: 22–28 °C (72–82 °F)
PREFERRED DIET: Live and dry foods
BREEDING: The paddle tip is assumed to perform a role in ensuring that the female adopts the appropriate orientation for effective sperm transfer. Eggs are fertilised internally and are later deposited. Hatching takes between 20–36 hours, depending on temperature
NOTES: The exact sperm transfer mechanism has not yet been established in this species. *C. riisei* meets the criteria for a livebearer as defined by John Wourms (1981) – see Part I for further discussion of this point

Plate 161   A young Swordtail Characin male, *Corynopoma riisei*. (*Mike Sandford*)

Index